The cardioprotective action of beta-blockers

The cardioprotective action of beta-blockers

Facts and theories

An international symposium held during the
7th European Cardiology Congress,
Amsterdam, 23rd June 1976

Edited by F. Gross, Heidelberg

University Park Press
Baltimore London Tokyo

Library of Congress Cataloging in Publication Data

Main entry under title:
The Cardioprotective action of beta-blockers.

1. Adrenergic beta receptor blockaders – Congresses. 2. Heart – Diseases – Chemotherapy – Congresses. I. Gross, Franz, 1913 – II. European Congress of Cardiology, 7th, Amsterdam, 1976.

RC684.A35C37   616.1'2'061   77-13243

ISBN 0-8391-1178-9

Published in North and South America by
University Park Press

ISBN 3-456-80491-1
© 1977 Hans Huber Publishers, Bern
Printed in Switzerland
by Basler Druck- und Verlagsanstalt, Basle

Contents

Introductory remarks
by F. Gross 9

The pharmacological basis for a cardioprotective action of beta-blockers
by H. Brunner 11
Summary of discussion 27

Beta-blockers and cardiac involvement in hypertension
by A. Zanchetti 28
Supplementary communications:
    Effects of beta-blockade on diurnal variability of blood pressure
    by W. H. Birkenhäger, P. W. de Leeuw, H. E. Falke, and A. Wester . 38
    The antihypertensive drug of first choice
    by F. R. Bühler 42
    Beta-blockade in hyperkinetic heart syndrome
    by M. Guazzi 46
Summary of discussion 49

Beta-blockers in the treatment of angina pectoris: the prevention of myocardial infarction
by J.-L. Rivier 52
Influence of beta-blockers on the incidence of re-infarction and sudden death after myocardial infarction
by C. Wilhelmsson, A. Vedin, and L. Wilhelmsen 58
Supplementary communications:
    Circulatory changes after beta-blockade in patients with coronary artery disease
    by A. Reale and M. Motolese 66
    Influence of long-acting nitrates and oxprenolol on rehabilitation after acute myocardial infarction
    by P. Rossi, A. Tamitz, A. Giordano, and G. Minuco 72
    The release and uptake of catecholamines by the myocardium in patients with coronary artery disease
    by W. Kübler and W. Mäurer 79

Clinical aspects of the treatment of cardiovascular diseases with beta-blocking drugs
by S. H. Taylor 81

Summary of concluding discussion
    Prepared contributions:
    B. GARNIER   93
    Z. MATKOVIĆ   94
    S. J. SARNOFF   96
    General discussion   96

Closing remarks
by F. GROSS   98

List of participants invited to present papers, supplementary communications, or discussion contributions

Prof. W. H. BIRKENHÄGER, Afdeling Inwendige Geneeskunde, Zuiderziekenhuis, Groeneveld 15, Rotterdam 3024, Netherlands

Prof. H. BRUNNER, Research Department (Pharmacology), Pharmaceuticals Division, CIBA-GEIGY LIMITED, Klybeckstrasse 141, CH-4002 Basle, Switzerland

Priv.-Doz. Dr. F. R. BÜHLER, Departement für Innere Medizin der Universität, Kantonsspital, Spitalstrasse 21, CH-4004 Basle, Switzerland

Priv.-Doz. Dr. B. GARNIER, rue St-Pierre 28, CH-1700 Fribourg, Switzerland

Prof. F. GROSS, Direktor des Pharmakologischen Institutes der Universität, Im Neuenheimer Feld 366, D-6900 Heidelberg, German Federal Republic

Prof. M. GUAZZI, Istituto di Ricerche Cardiovascolari dell'Università di Milano, Via F. Sforza 35, I-20122 Milan, Italy

Prof. W. KÜBLER, Abteilung Innere Medizin III (Kardiologie), Medizinische Universitätsklinik, Bergheimer Strasse 58, D-6900 Heidelberg, German Federal Republic

Dr. Z. MATKOVIĆ, Farmakološko-klinički Otsjek, Istraživačko Odjeljenje, Opća Bolnica "Dr. Josip Kajfeš", Pavleka Miškine 64, YU-41000 Zagreb, Yugoslavia

Prof. A. REALE, Direttore, Cattedra II Malattie Cardiovascolari, Università di Roma, Policlinico Umberto I, I-00161 Rome, Italy

Prof. J.-L. RIVIER, Médecin-Chef, Division de cardiologie du Département de médecine, Centre hospitalier universitaire vaudois, rue du Bugnon 17, CH-1011 Lausanne, Switzerland

Prof. P. ROSSI, Primario, Divisione di Cardiologia, Ospedale Maggiore della Carità, Corso Mazzini 18, I-28100 Novara, Italy

Dr. S. J. SARNOFF, M.D., F.A.C.C., Bethesda, Md., U.S.A.

Dr. S. H. TAYLOR, B.Sc., M.B., Ch.B., F.R.C.P. (Edin.), University Department of Cardiovascular Studies and Department of Medical Cardiology, The Martin Wing, The General Infirmary, Leeds LS1 3EX, England

Dr. C. WILHELMSSON, Medicinska kliniken I, Sahlgrenska sjukhuset, S-41345 Gothenburg, Sweden

Prof. A. ZANCHETTI, Direttore, Istituto di Patologia Medica I dell'Università di Milano, Via F. Sforza 35, I-20122 Milan, Italy

Introductory remarks

by F. Gross*

There is little if any doubt not only that the beta-adrenergic blocking drugs mark the most significant advance made in the treatment of cardiovascular disorders during the past decade, but also that, broadly speaking, they constitute one of the few active therapeutic principles currently available for use in cardiology. So attractive are their properties, in fact, that the beta-blockers – or beta-sympatholytics – have certainly not escaped the danger of inducing exaggerations and overstatements. Of this danger one of the consequences that we are witnessing today is the ever-increasing number of similar drugs which differ from one another at best only marginally and none of which has any really significant advantages to offer. About a dozen different beta-blockers have already been marketed in various countries, and more are due to follow shortly. The opposite trend can be observed in the U.S.A., where the veteran propranolol is still the only beta-blocker on the market and has not yet been accepted by the Food and Drug Administration for the treatment of high blood pressure**. It is difficult to say which situation is to be preferred or with which it is easier to live.

The other problem with successful new drugs is that too much may be expected from them, and that manufacturers as well as eager clinicians are quite often tempted to make excessive claims for them and to regard them with insufficient criticism and caution. The beta-blockers have now acquired a firm place in the treatment of various forms of cardiac arrhythmia, coronary heart disease, and hypertension – i.e. three groups of cardiovascular disorders which are well defined, even though they may be quite different in their aetiology and pathogenesis. The present symposium, however, is concerned with the cardioprotective action of the beta-blockers; in contrast to the diseases I have just mentioned, we shall thus be dealing, not with clearly defined clinical entities, but with a concept which lends itself to a variety of interpretations.

Cardioprotection is a simplifying slogan. To protect the heart sounds fine, but against what are we trying to protect it? Against damage, against dysfunction, against overloading, or against toxic effects? There are numerous noxious influences which may act on the heart, and it is only against some of these that the beta-blockers – thanks especially to their ability to reduce the work load and the oxygen consumption of the heart – are capable of affording protection. Hence, when we speak of cardioprotection, we must be aware of the fact that this term is used for the sake of simplicity and that it covers various possible effects which the beta-blockers may exert on the heart and does not imply that this class of drugs has a clearly defined pattern of "cardioprotective" activity. Efforts should nevertheless be made to standardise the criteria for such cardioprotective activity and to clarify precisely in which way and by which effects in particular the beta-blockers may be able to protect the heart from further damage or from severe disorders of cardiac rhythm resulting in sudden death. At the same time, it

---

* Pharmakologisches Institut der Universität, Heidelberg, German Federal Republic.
** Shortly after this symposium had been held, propranolol was also approved by the F.D.A. for use in hypertension.

should also be borne in mind that beta-blockers involve the inherent risk of inducing heart failure – especially in elderly patients, in whom enhanced sympathetic drive may act as a compensatory mechanism serving to maintain the requisite cardiac index. Not all the effects of beta-blockade do in fact contribute to the protection of the heart.

During our symposium, reference will be made to various possibilities of cardioprotection by means of beta-blockers, to the pharmacological basis of this protection, and to its practical clinical implications. I hope that our meeting will not only help to indicate more clearly where we stand today but also assist us in deciding how to proceed, and whither to proceed, in the future. In my opinion, we have now reached a stage at which we should try to draw conclusions from the clinical data available and make recommendations as to their application in practice, and it is with these aims in view that I would suggest our proceedings should be conducted.

The pharmacological basis for a cardioprotective action of beta-blockers

by H. Brunner*

Strictly speaking, the cardioprotective action of the beta-blockers as described in clinical studies entails two separate, though closely interrelated effects:

1. A reduction in the incidence of myocardial infarction or, in the presence of an infarction, a reduction in the size of the infarct.
2. A reduction in the incidence of sudden death resulting from severe cardiac arrhythmias.

The principal mechanisms which, in various combinations and permutations, might be responsible for these two effects are as follows:

1. Diminution in myocardial oxygen consumption.
2. Antagonism of locally and systemically released catecholamines.
3. Anti-arrhythmic action.
4. Influences on myocardial energy and possibly also mineral metabolism, reduction in blood viscosity, etc.

Before dealing with these various mechanisms in detail, it is important to make certain reservations:

Firstly, treatment with beta-blockers represents only one of several methods of minimising the damage caused by a fresh infarction. The same objective can also be achieved by a number of other measures, such as, for example: balloon counterpulsation; infusions of glucose, insulin, and potassium in order to facilitate anaerobic glycolysis; and infusions of hyaluronidase, so as to increase oxygen diffusion[4]. Nitroglycerin, too, has the effect of reducing the size of a fresh infarction, provided a marked fall in blood pressure can be prevented[14]. In contrast to the beta-blockers, however, none of these measures is employed on a large scale in the management of hypertension (an indication in which the prevention of cardiac complications is so important), because their value is to some extent disputed.

Secondly, beta-blockers can have dangerous consequences in cases of fresh infarction and make the patient's situation even worse. They may, for example, so impair impulse formation and conduction that cardiac arrest ensues. Furthermore, by eliminating the positive inotropic effect of the catecholamines, they may aggravate heart failure. In cases of cardiogenic shock, moreover, they are strictly contra-indicated.

The mechanisms by which the beta-blockers exert a cardioprotective action can be more readily understood by considering the most important processes involved in the development of an infarction and by examining the question of how these processes can be influenced by beta-blockers.

---

* Research Department (Pharmacology), Pharmaceuticals Division, CIBA-GEIGY LIMITED, Basle, Switzerland.

Processes involved in the development of an infarction

In the first few hours after an acute ischaemic episode the myocardium displays a degree of hypoxia which may vary depending on the opening up of collaterals, the formation or resolution of thrombi, the occurrence of changes in perfusion pressure and ventricular performance, etc. Following experimental, i.e. complete, occlusion of a coronary artery, at least two main hypoxic zones can be distinguished. The first zone, marked by extremely severe hypoxia (due to a decrease in blood flow to less than 10% of control values) and by the presence of necrotising tissue, is no longer capable of contracting actively within a matter of seconds after the arterial occlusion, while later on it is passively stretched during systole[41]. This zone is surrounded by a second zone, in which the degree of hypoxia decreases towards the borderline with normal myocardial tissue, i.e. as the collateral blood supply improves. This second zone displays reduced contractility and can also be distinguished from normal myocardial tissue histologically, e.g. by dehydrogenase staining[9]. The size of the infarct and of the surrounding area of hypoxia does not become finally fixed until several hours after the occurrence of the arterial occlusion. In experiments conducted by Cox et al.[9] (Figure 1), for example, the damage did not reach its ultimate extent until 18 hours had passed. As far as time is concerned, therefore, a good opportunity exists of influencing the size of a fresh infarct.

In the ischaemic myocardium a number of biochemical changes occur: the energy-rich phosphates (adenosine triphosphate and creatine phosphate) disappear very rapidly, while anaerobic glycolysis is enhanced, leading to an increase in lactate and other acid products of metabolism. The resultant acidosis already markedly impairs myocardial function[3]. What is more, the accumulation of osmotically active particles

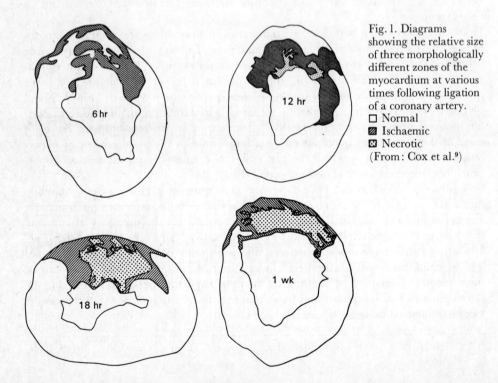

Fig. 1. Diagrams showing the relative size of three morphologically different zones of the myocardium at various times following ligation of a coronary artery.
☐ Normal
▨ Ischaemic
▩ Necrotic
(From: Cox et al.[9])

gives rise to swellings which impede the diffusion of oxygen. Finally, potassium escapes from the damaged cells, with the result that the surrounding cells are exposed to high extracellular concentrations of potassium which reduce their transmembrane resting potential and increase the risk of arrhythmias[10, 22].

As regards electrophysiological parameters, hypoxia *per se* produces first of all a shortening of the action potential. Not until the hypoxia has been present for some considerable time do the resting and action potentials decrease, while the duration of the rise in the action potential is prolonged and the conduction velocity diminishes[52]. In the heart as a whole, it is above all the shortening of the action potential (i.e. the reduction in the plateau portion of the curve)[7], in addition to the lowering of the stimulus threshold during the relative refractory period (vulnerable period)[57], that is probably of essential importance in the triggering of fibrillation.

During the development of an infarction a major role is played by catecholamines. In the presence of anoxia approximately one-quarter of the noradrenaline contained in the myocardium is released within only a few minutes (in the rabbit heart, for example, the amount of noradrenaline thus released averages 0.46 mcg./g.[59]); a high local concentration of noradrenaline therefore has to be reckoned with. Furthermore, especially in patients with anterior infarction[56], general sympathetic tone is increased – partly no doubt by means of a reflex-induced mechanism and partly in response to concomitant anxiety and pain – and this increase is accompanied by a marked rise in the blood levels of adrenaline and noradrenaline[26, 27, 54]. This sympatho-adrenal stimulation, which may persist for days in severe cases, has considerable repercussions on the myocardial tissue in the hypoxic marginal zone. Since the cells in this zone are still perfectly capable of responding to positive inotropic stimuli, their oxygen requirement increases, so that both the hypoxia and the resultant damage are aggravated[33].

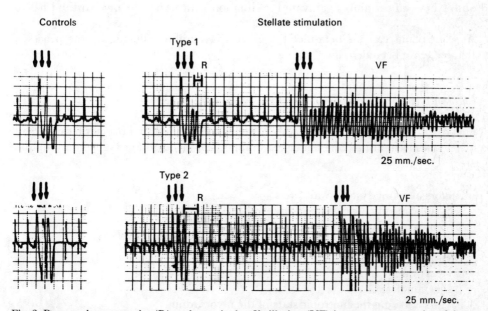

Fig. 2. Repeated extrasystoles (R) and ventricular fibrillation (VF) in response to a series of three artificial impulses (R/T type) at twice the threshold voltage (↓↓↓) in the dog; in the upper example, the repeated extrasystoles follow immediately upon the R/T impulses, while in the lower one they set in after a certain interval. In the absence of stellate stimulation (controls) only the three artificially induced extrasystoles are observed. (From: VERRIER et al.[53])

Catecholamines also give rise in the myocardium to a number of electrophysiological changes which add to the disturbances provoked by hypoxia and predispose the muscle to fibrillation: they accentuate the slope of spontaneous phase IV depolarisation of automatic cells and Purkinje fibres and thus induce increased firing. Since, however, different cells display different degrees of sensitivity, new pacemakers emerge, conduction may become disorganised, and multiple re-entry paths may appear, with the result that eventually fibrillation can occur[25]. *In situ*, these phenomena are reflected in an increase in automaticity, in a decrease in diastolic stimulation threshold and duration of the refractory period, in non-uniformity of recovery of excitation, and in a lower fibrillation threshold[8, 20].

In dogs, for example, it has been shown that sequential pulsing of a normal ventricle with stimuli at only twice the threshold level, which would normally induce only one extrasystole on each occasion, can lead to ventricular fibrillation in 60% of the animals if the stellate ganglion is stimulated at the same time[53] (Figure 2). Conversely, it has been demonstrated by a number of investigators that chronic denervation of the heart, accompanied by depletion of the myocardial catecholamine stores, affords protection against ventricular fibrillation[13, 21, 46]. In addition, an increase in systemic sympathetic tone produces a number of metabolic changes, of which the most important are the following: in the liver, it leads not only to glycogenolysis which, together with inhibition of insulin secretion, can provoke a considerable rise in blood sugar levels, but also to lipolysis accompanied by an increase in the supply of free fatty acids to the heart[36] – and high free fatty acid levels increase the oxygen consumption of the myocardium without improving its mechanical efficiency[34]. Furthermore, there is evidence suggesting that the incidence of arrhythmias following infarction increases in the presence of a very high serum level of free fatty acids[37]. Although the importance of these two factors cannot be so clearly assessed as that of the others mentioned previously, they can certainly aggravate the situation with which the myocardium has to cope.

At what points can the beta-blockers exert a favourable influence on the processes that have just been described?

Effects of beta-blockers on the development and size of an infarction

First of all, the beta-blockers are able to reduce the myocardial oxygen requirement by several mechanisms. The major determinants of myocardial oxygen consumption are:

– Factors affecting systolic wall stress:
    Systolic intraventricular pressure
    Ventricular size
    Configuration and thickness of the ventricular wall
– Factors affecting duration of systolic wall stress:
    Heart rate
    Ejection time
– Factors affecting the inotropic state of the myocardium:
    e.g. catecholamines.

While the beta-blockers influence all these factors, this influence does not invariably entail a reduction in oxygen consumption.

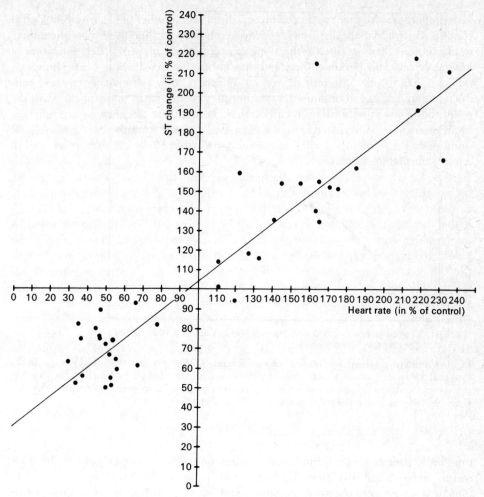

Fig. 3. Effects of various heart rates, in the presence of a normal arterial blood pressure, on the extent of ST elevation five minutes after coronary occlusion. (From: REDWOOD et al.[42])

1. Beta-blockers lower systolic intraventricular pressure to the same extent as they reduce arterial blood pressure, and this extent, especially in hypertensives, is quite appreciable. Theoretically, a prolonged reduction in systolic blood pressure, as well as the resultant easing of the strain on the heart, could also be accompanied by a decrease in left-ventricular hypertrophy, which would be an additional benefit. In principle, however, these effects are likely to be produced by any antihypertensive treatment and are not specific for beta-blockers[48].

2. Beta-blockers appreciably reduce heart rate, particularly in the presence of increased sympathetic tone, i.e. during physical exercise or emotional stress.
The reduction in heart rate is probably one of the essential factors responsible for a saving in oxygen. In studies performed on conscious dogs in which the heart rate was varied between 30 and 210 beats/minute (by means of vagal stimulation or the use of an artificial pacemaker), REDWOOD et al.[42] found that, following occlusion of a coronary artery, the sum of the ST elevations in the E.C.G. recorded intramyocardially

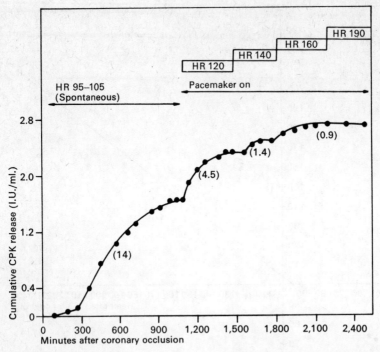

Fig. 4. Cumulative release of creatine phosphokinase (CPK) as heart rate (HR) is gradually increased following coronary occlusion in conscious dogs; the figures in brackets indicate the calculated size of the myocardial necrosis (expressed in gramme equivalents of creatine phosphokinase) corresponding to the given heart rate. (From: SHELL and SOBEL[47])

from the ischaemic area was linearly dependent on the heart rate (Figure 3). In other words, there is a linear relationship between heart rate and cell damage.

Similar findings were obtained by SHELL and SOBEL[47], although these authors employed a completely different method of estimating the size of the ischaemic area, i.e. cumulative determination of the creatine phosphokinase released from the damaged heart cells. Since creatine phosphokinase – especially the BM (i.e. myocardium-specific) iso-enzyme – is only released into the blood when myocardial cells are actually destroyed, the extent of the release makes it possible to draw conclusions about the size of the infarction. Figure 4 shows how, following ligature of a coronary artery in dogs, the amount of enzyme released gradually increases when the heart rate is stepped up by means of a pacemaker.

3. Beta-blockers inhibit the positive inotropic effect of the catecholamines on the heart. As regards the administration of beta-blockers for the prevention of infarction, the following point should be borne in mind: as long as the patient remains supine and relaxed, and therefore displays hardly any sympathetic tone, beta-blockade will produce virtually no perceptible effect. Its effect will only become noticeable when the patient stands up or performs physical work. The higher the level of sympathetic tone in a given situation, the more pronounced the effect of a beta-blocker. Peak loads on the myocardium, which of course invariably represent peaks in oxygen consumption, are suppressed, with the result that the risk of an infarction in patients with a limited coronary reserve is reduced.

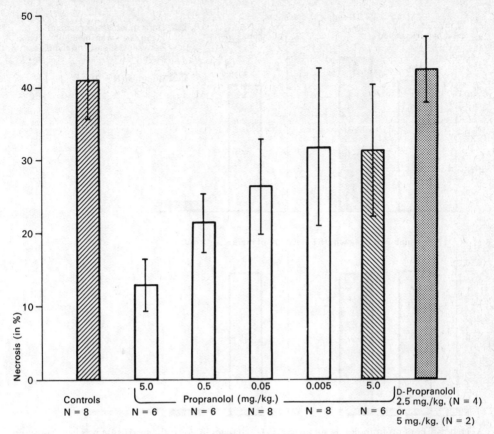

Fig. 5. Extent of necrosis in the posterior papillary muscle (in percent) 40 minutes after temporary occlusion of the left circumflex coronary artery in dogs. (Adapted from: REIMER et al.[43])

▨ Controls
☐ Propranolol, 10 minutes *before* occlusion
▧ Propranolol, 30 minutes *after* occlusion
▦ D-Propranolol, 10 minutes *before* occlusion

The same applies to the cells in the hypoxic marginal zone of an infarction: here, too, beta-blockers inhibit the positive inotropic effect of the catecholamines, achieve a saving in oxygen, and thus possibly enable the cells to survive.

4. In addition, however, beta-blockers also give rise to changes which lead to an increase in myocardial oxygen requirement: in response to beta-blockers ventricular size usually increases and ejection time is prolonged. But these effects are less pronounced. On the whole, the net result of the beta-blockers' influence on myocardial oxygen requirement is to reduce that requirement. None the less, the importance of the individual factors may of course vary considerably from case to case.

The question of whether the beta-blockers can prevent infarction, or reduce the size of an infarct, has been intensively studied by several teams of investigators in animal experiments.

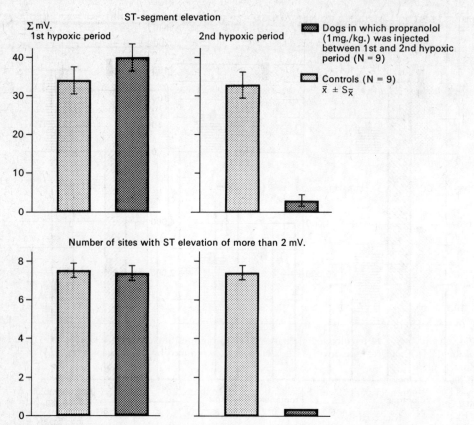

Fig. 6. Effect of propranolol on hypoxic E.C.G. changes in dogs; the heart rate was kept constant at 162 ± 12 beats/minute, and the mean arterial pressure at 110 ± 7 mm. Hg. (From: ASKENAZI and MAROKO[1])

In one series of such experiments SOMMERS and JENNINGS[49] investigated the effect of propranolol on the necrosis occurring 20–25 minutes after transient ischaemia had been induced in dogs by clamping a branch of the coronary artery. Among the animals that survived a coronary artery ligature lasting 20 minutes, eight out of 15 controls displayed infarction, whereas infarction was found in only two of the 14 animals that had been given propranolol by intra-arterial injection in a dosage of 5 mg./kg. prior to the transient ischaemia. In cases where the coronary blood flow was occluded for 25 minutes, the infarction rate was 11 out of 16 for the controls and one out of 16 for the animals given the beta-blocker. Heart rate was reduced in the treated group (111 ± 7 beats/minute, as against 140 ± 6 beats/minute in the controls), and systolic blood pressure, too, was lower in the animals under beta-blockade (123 ± 6 mm. Hg) than in the controls (138 ± 5 mm. Hg).

In analogous experiments carried out in dogs by the same research team[43], blood flow in the circumflex coronary artery was occluded for 40 minutes in order to create a situation in which infarction would occur in all the control animals. Propranolol was administered intravenously in different doses either ten minutes prior to coronary artery occlusion or – in the highest dose used – 30 minutes after the occlusion. Among the animals that survived for 2–5 days after the occlusion the amount of necrotic tissue

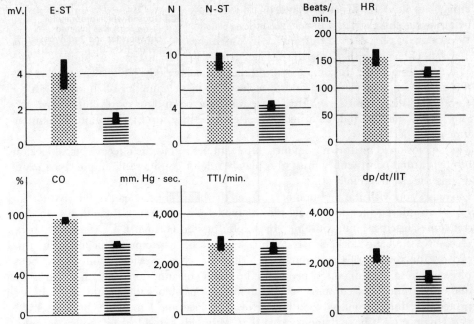

Fig. 7. Comparison of haemodynamic parameters 15 minutes after artificial coronary occlusion in the dog without (⋮⋮⋮) and with (≡) beta-blockade induced by propranolol. (From: WATANABE et al.[55])

E-ST   = mean ST elevation
N-ST   = number of sites with ST elevation of more than 2 mV.
HR   = heart rate
CO   = cardiac output
TTI   = tension-time index (left ventricle)
dp/dt/IIT = index of rise in pressure (left ventricle)

in the posterior papillary muscle, expressed in percent, was found to have decreased in a dose-dependent manner. D-Propranolol, the isomer lacking beta-blocking activity, proved inactive. Treatment given 30 minutes after the occlusion likewise exerted hardly any influence on the size of the infarct (Figure 5).

THEROUX et al.[51] report that loss of function in the hypoxic marginal zone was lower following treatment with propranolol than in control experiments in which the dogs received no medication.

MAROKO et al.[33], ASKENAZI and MAROKO[1], and WATANABE et al.[55] studied the effect of beta-blockade on the size of the ischaemic area with the aid of surface mapping. Of the findings obtained in these studies two deserve particular attention inasmuch as they shed new light on the subject: in dogs subjected to constriction of the descending branch of the anterior coronary artery and to artificial respiration with 10% oxygen for 15 minutes, ASKENAZI and MAROKO found that ST-segment elevations were clearly reduced following treatment with propranolol in a dose of 1 mg./kg. and that the number of sites with an ST elevation of more than 2 mV. decreased to zero (Figure 6; the corresponding control values, derived from a comparable group of dogs exposed to hypoxia on two occasions without treatment, are also shown in this figure). One important point to be borne in mind is that in these studies the heart rate and the mean arterial blood pressure were kept constant by artificial means. Hence, even though its effect on heart rate and on blood pressure was thereby eliminated, the beta-blocker still exerted a beneficial action.

WATANABE et al.[55] carried out similar investigations in which they analysed haemodynamic changes at the same time. Their results are summarised in Figure 7: once again, beta-blockade – in comparison with the control period – led to a clear-cut reduction in ST-segment elevation and to a decrease in the number of sites with an ST elevation of more than 2 mV. Heart rate was reduced in the treated animals, as were also cardiac output and, as a measure of contractility, dp/dt/IIT, but not tension-time index per minute.

The conclusion to be drawn from these latter studies is that the effect of beta-blockade on myocardial contractility must also play a role of considerable importance in reducing the size of an infarct.

In connection with the question of whether the beta-blockers can in fact reduce the size of an infarct, mention should also be made of one effect of these drugs which is sometimes listed as unfavourable: by inhibiting the coronary, i.e. vascular, beta-receptors, the beta-blockers may "unmask" a vasoconstrictor alpha tone and thus provoke narrowing of the coronary arteries. Since, as already pointed out, beta-blockers cause a reduction in myocardial oxygen consumption, and since this is the main determinant of coronary artery diameter, beta-blockade should really be expected to reduce coronary blood flow. However, PITT and CRAVEN[40], BECKER et al.[2], and HILLIS et al.[23] have demonstrated that, although beta-blockers reduce coronary blood flow in the non-hypoxic regions of the heart, they do not do so in the ischaemic area. Hypoxia is apparently such a powerful stimulus to coronary artery dilatation that the vessels in the hypoxic area no longer respond to weak constrictor stimuli. Consequently, beta-blockers give rise to a redistribution of the coronary blood flow. In addition, two groups of investigators, using different methods, have shown that beta-blockers improve perfusion in the subendocardial layers of the left ventricle at the expense of subepicardial perfusion[2,19]. It is, however, precisely the subendocardial layers that are particularly at risk from infarction.

Beta-blockers can also exert an indirect effect on the hypoxic myocardium by inhibiting certain systemic repercussions of the increased sympathetic tone prevailing in the post-infarction phase. The chief repercussion to be mentioned in this connection is lipolysis. Beta-blockers reduce the rise in blood free fatty acid levels, thereby lessening ischaemic damage to the myocardium, as well as eliminating the possible arrhythmogenic effect of the free fatty acids and, at the same time, preventing them from accumulating in the hypoxic tissue.

Not yet elucidated is the significance of findings indicating that beta-blockers inhibit platelet aggregation[5] and diminish blood viscosity[11]. Theoretically, a reduction in the tendency to thrombosis and a lower blood viscosity would also have a cardioprotective action. To date, however, there is only scanty evidence to show that beta-blockers might likewise produce an antithrombotic effect in clinical use[17,45].

Effects of beta-blockers in diminishing the incidence of sudden death

An essential feature of the cardioprotective action of the beta-blockers is that, besides preventing infarction or reducing the size of an infarct, they are also able to lower the incidence of sudden death.

Sudden death must be regarded as a problem in its own right, because approximately two-thirds of all deaths occurring in association with coronary heart disease are

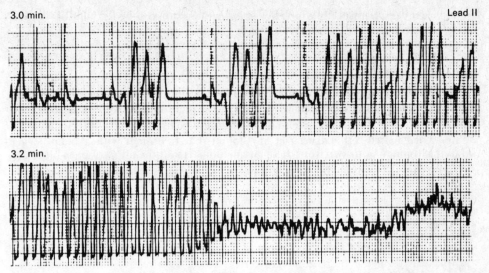

Fig. 8. Ventricular arrhythmias following balloon occlusion of the descending branch of the left coronary artery in the dog. In the first three minutes after coronary occlusion several brief successive bursts of ventricular tachycardia occurred. A few seconds later "ventricular tachycardia in the vulnerable period" set in and developed into ventricular fibrillation. (From: LOWN and WOLF[32])

"sudden"[29], and only one-quarter of them occur in hospital[18]. Sudden death is in all probability due to a disturbance of cardiac rhythm which begins with short bursts of ventricular tachycardia and which eventually develops into ventricular fibrillation. By way of an example, the pattern of events occurring in a dog three minutes after occlusion of a coronary artery is shown in Figure 8[32]. Though the causes responsible for such arrhythmia are complex, it is the combination of hypoxia and a high catecholamine concentration in the marginal zone that most probably plays a cardinal role in its development[31].

Since arrhythmias induced by ligation of a coronary artery are commonly employed as a pharmacological model by which to investigate the anti-arrhythmic activity of beta-blockers, numerous animal-experimental findings have been published indicating that beta-blockers can afford the heart muscle protection against arrhythmias of this type. A few of these findings will be mentioned here by way of illustration.

KHAN et al.[28] reported that fatal ventricular fibrillation set in within 14 minutes after occlusion of a coronary artery in 18 out of 25 dogs under control conditions. This number was reduced to six out of 25 when propranolol was administered in a dose of 0.1 mg./kg. i.v., whereas D-propranolol, which is ineffective as a beta-blocker, exerted no influence on the mortality rate. It is interesting to note that in these experiments a higher dose of propranolol – namely, 1.0 mg./kg. i.v. – was less effective than the lower dose. In response to the higher dose, moreover, half of the animals died not of fibrillation but of cardiac standstill. Basically similar findings, including in some instances a less good effect with higher doses, were obtained by PENTECOST and AUSTEN[39] with propranolol, and by other groups of investigators with various beta-blockers (sotalol in the above-mentioned study by KHAN et al.[28], and nadolol in an experiment reported by EVANS et al.[15]). It must thus be concluded that here dosage is critical and should therefore be adjusted with great care.

What are the mechanisms by which beta-blockers can suppress ventricular fibrillation induced by occlusion of a coronary artery?

1. Reduction in the size of the hypoxic area, a mechanism to which reference has already been made.

2. Inhibition of direct catecholamine effects in this area, which is certainly the most important mechanism, especially as it raises the fibrillation threshold[58]; prolongation of the duration of the action potential; in dogs in which a coronary artery had been temporarily ligated, KUPERSMITH et al.[30] found that treatment with a beta-blocker slowed impulse conduction in the ischaemic zone, besides prolonging the effective refractory period and the duration of the action potential (the disparity between the

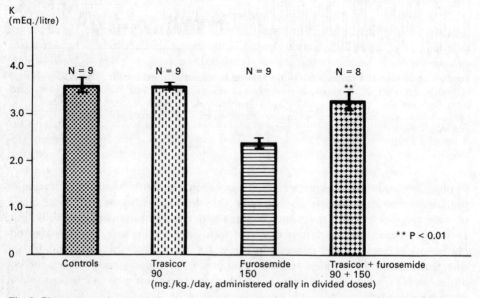

Fig. 9. Plasma potassium levels in the rat following four days' oral treatment with Trasicor (90 mg./kg. daily), furosemide (150 mg./kg. daily), and Trasicor together with furosemide in the same dosages.

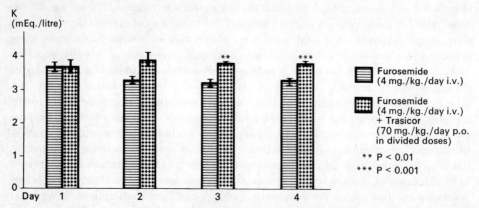

Fig. 10. Plasma potassium levels in the dog following treatment with furosemide alone and with furosemide plus Trasicor (N = 3).

duration of the action potential in normal and hypoxic tissue was also diminished). These mechanisms, which serve to reduce abnormal excitation, provide a convincing explanation for the anti-arrhythmic effect of beta-blockade.

3. Possibly an indirect effect exerted via a reduction in the rise of serum free fatty acids.

4. A membrane-stabilising effect similar to that of local anaesthetics or of quinidine. This effect, however, is exerted only by some beta-blockers, such as propranolol, oxprenolol (®Trasicor), and alprenolol (®Gubernal), which reduce upstroke velocity and overshooting of the action potential, diminish the highest stimulation frequency which the heart can follow, and exercise similar influences that are capable of normalising disturbances of cardiac rhythm[16, 38].

Mention should also be made of another effect which can occur only in hypertensives receiving combined therapy with a beta-blocker and a diuretic. Thanks to their efficacy and good tolerability, combinations of this type are of course being widely employed nowadays and are recommended in some quarters as the method of choice for the management of moderately severe hypertension. Diuretics may give rise to hypokalaemia as a side effect. This hypokalaemia is due largely to the secondary hyperaldosteronism with which the organism endeavours to combat the loss of sodium caused by the diuretic. One important factor in the control of aldosterone secretion is the renin-angiotensin system. The administration of a diuretic is followed by an increase in the amount of renin released by the kidney. This increase, via a rise in angiotensin production, helps to trigger off secondary hyperaldosteronism. Beta-blockers, however, are able to inhibit renin production[6]. Consequently, their use is likely to lead to a decrease in the production of angiotensin and also to a reduction in the degree of hyperaldosteronism. As a result, less potassium should be lost than in response to the diuretic alone. This has, in fact, been clearly demonstrated in animal experiments:

If rats are treated with large doses of furosemide for four days, their plasma potassium levels will be appreciably reduced. This effect, however, can be prevented by administering oxprenolol at the same time (Figure 9). The same applies to dogs (Figure 10). It should be emphasised in this connection that neither diuresis nor sodium excretion is affected by additional administration of the beta-blocker. Similar results have been obtained by Sweet and Gaul[50] using a different diuretic and a different beta-blocker (timolol).

Findings comparable to those recorded in animal experiments have also been obtained in patients. In an Italian multicentre study involving 200 patients, for example, hypokalaemia (plasma potassium levels of less than 3.5 mEq./litre) occurred in 5% of cases when oxprenolol in a daily dosage of 160–320 mg. was administered in combination with 40 mg. chlortalidone daily; on the other hand, hypokalaemia was encountered in 11% of cases when the same daily dosage of chlortalidone was given as monotherapy for six weeks.

The potassium-saving effect of the beta-blockers, which, according to the results of investigations undertaken by Drayer et al.[12], may be equivalent to that of spironolactone, is of course quite likely to play a role in reducing the risk of cardiac complications during treatment for hypertension.

To sum up, it is safe to say that beta-blockers can reduce both the incidence of myocardial infarction and the size of an infarct. They do this via several mechanisms, the most important of which are a reduction in heart rate, a diminution in ventricular systolic wall stress, and elimination of local and systemic effects of catecholamines. The decisive component in all these actions of the beta-blockers is probably the reduction in myocardial oxygen consumption, or at least in its peak values. In addition, beta-blockers can reduce the incidence of ventricular fibrillation following occlusion of a coronary artery. While this effect is mediated via the same mechanisms as those responsible for preventing infarction or reducing the size of an infarct, it must be borne in mind that certain beta-blockers also possess anti-arrhythmic properties. In some cases, additional effects of the beta-blockers, such as inhibition of platelet aggregation, or the potassium-saving effect achieved by prolonged administration of a beta-blocker together with a diuretic, may also play a role.

Although the importance of each individual factor will of course depend on the situation of the patient concerned, it is the very plurality of the effects exerted by the beta-blockers that accounts for the clear-cut cardioprotective action of these drugs.

References

1 ASKENAZI, J., MAROKO, P.R.: Prevention of hypoxia-induced myocardial damage by propranolol in dogs with partial coronary occlusions. Clin. Res. *23*, 561 A (1975); abstr.
2 BECKER, L.C., FORTUIN, N.J., PITT, B.: Effect of ischemia and antianginal drugs on the distribution of radioactive microspheres in the canine left ventricle. Circulat. Res. *28*, 263 (1971)
3 BRACHFELD, N.: The glucose-insulin-potassium (GIK) regimen in the treatment of myocardial ischemia. Circulation *48*, 459 (1973)
4 BRAUNWALD, E., MAROKO, P.R., LIBBY, P.: Reduction of infarct size following coronary occlusion. Circulat. Res. *34/35*, Suppl. III: 192 (1974)
5 BUCHER, H.W., STUCKI, P.: The effect of various beta-receptor blocking agents on platelet aggregation. Experientia (Basle) *25*, 280 (1969)
6 BÜHLER, F.R., LARAGH, J.H., VAUGHAN, E.D., Jr., BRUNNER, H.R., GAVRAS, H., BAER, L.: Antihypertensive action of propranolol. Specific antirenin responses in high and normal renin forms of essential, renal, renovascular and malignant hypertension. Amer. J. Cardiol. *32*, 511 (1973)
7 BURN, J.H., HUKOVIĆ, S.: Anoxia and ventricular fibrillation; with a summary of evidence on the cause of fibrillation. Brit. J. Pharmacol. *15*, 67 (1960)
8 CEREMUŻIŃSKI, L., STASZEWSKA-BARCZAK, J., HERBACZYNSKA-CEDRO, K.: Cardiac rhythm disturbances and the release of catecholamines after acute coronary occlusion in dogs. Cardiovasc. Res. *3*, 190 (1969)
9 COX, J.L., MCLAUGHLIN, V.W., FLOWERS, N.C., HORAN, L.G.: The ischemic zone surrounding acute myocardial infarction. Its morphology as detected by dehydrogenase staining. Amer. Heart J. *76*, 650 (1968)
10 CUMMINGS, J.R.: Electrolyte changes in heart tissue and coronary arterial and venous plasma following coronary occlusion. Circulat. Res. *8*, 865 (1960)
11 DINTENFASS, L., LAKE, B.: Beta blockers and blood viscosity. Lancet *1976/I*, 1026; corresp.
12 DRAYER, J.I.M., KLOPPENBORG, P.W.C., FESTEN, J., LAAR, A. VAN'T, BENRAAD, T.J.: Intrapatient comparison of treatment with chlorthalidone, spironolactone and propranolol in normoreninemic essential hypertension. Amer. J. Cardiol. *36*, 716 (1975)
13 EBERT, P.A., VANDERBEEK, R.B., ALLGOOD, R.J., SABISTON, D.C., Jr.: Effect of chronic cardiac denervation on arrhythmias after coronary artery ligation. Cardiovasc. Res. *4*, 141 (1970)
14 EPSTEIN, S.E., KENT, K.M., GOLDSTEIN, R.E., BORER, J.S., REDWOOD, D.R.: Reduction of ischemic injury by nitroglycerin during acute myocardial infarction. New Engl. J. Med. *292*, 29 (1975)
15 EVANS, D.B., PESCHKA, M.T., LEE, R.J., LAFFAN, R.J.: Anti-arrhythmic action of nadolol, a β-adrenergic receptor blocking agent. Europ. J. Pharmacol. *35*, 17 (1976)
16 FREEMAN, S.E., TURNER, R.J.: The effects of l-propranolol and practolol on atrial and nodal transmembrane potentials. J. Pharmacol. exp. Ther. *195*, 133 (1975)
17 FRISHMAN, W.H., WEKSLER, B., CHRISTODOULOU, J.P., SMITHEN, C., KILLIP, T.: Reversal of abnormal platelet aggregability and change in exercise tolerance in patients with angina pectoris following oral propranolol. Circulation *50*, 887 (1974)
18 FULTON, M., JULIAN, D.G., OLIVER, M.F.: Sudden death and myocardial infarction. Circulation *39/40*, Suppl. IV: 182 (1969)

19 GROSS, G.J., WINBURY, M.M.: Beta adrenergic blockade on intramyocardial distribution of coronary blood flow. J. Pharmacol. exp. Ther. *187*, 451 (1973)
20 HAN, J., GARCIA DE JALON, P., MOE, G.K.: Adrenergic effects on ventricular vulnerability. Circulat. Res. *14*, 516 (1964)
21 HARRIS, A.S., BISTENI, A., RUSSELL, R.A., BRIGHAM, J.C., FIRESTONE, J.E.: Excitatory factors in ventricular tachycardia resulting from myocardial ischemia. Potassium, a major excitant. Science *119*, 200 (1954)
22 HARRIS, A.S., ESTANDIA, A., TILLOTSON, R.F.: Ventricular ectopic rhythms and ventricular fibrillation following cardiac sympathectomy and coronary occlusion. Amer. J. Physiol. *165*, 505 (1951)
23 HILLIS, W.S., TAYLOR, K.M., CONELY, J., LAWRIE, T.D.V., HUTTON, I.: Protection of the ischaemic myocardium by propranolol. In: VIIth Europ. Congr. Cardiol., Amsterdam 1976, Abstr. Book I, p. 56 *et seq.*
24 HOAK, J.C., WARNER, E.D., CONNOR, W.E.: Platelets, fatty acids and thrombosis. Circulat. Res. *20*, 11 (1967)
25 HOFFMAN, B.F., SINGER, D.H.: Appraisal of the effects of catecholamines on cardiac electrical activity. Ann. N.Y. Acad. Sci. *139*, 914 (1967)
26 JÉQUIER, E., PERRET, C.: Urinary excretion of catecholamines and their main metabolites after myocardial infarction; relationship to the clinical syndrome. Europ. J. clin. Invest. *1*, 77 (1970)
27 JEWITT, D.E., MERCER, C.J., REID, D., VALORI, C., THOMAS, M., SHILLINGFORD, J.P.: Free noradrenaline and adrenaline excretion in relation to the development of cardiac arrhythmias and heart-failure in patients with acute myocardial infarction. Lancet *1969/I*, 635
28 KHAN, M.I., HAMILTON, J.T., MANNING, G.W.: Protective effect of beta adrenoceptor blockade in experimental coronary occlusion in conscious dogs. Amer. J. Cardiol. *30*, 832 (1972)
29 KULLER, L., COOPER, M., PERPER, J.: Epidemiology of sudden death. Arch. intern. Med. *129*, 741 (1972)
30 KUPERSMITH, J., SHIANG, H., LITWAK, R.S., HERMAN, M.V.: Electrophysiological and anti-arrhythmic effects of propranolol in canine acute myocardial ischemia. Circulat. Res. *38*, 302 (1976)
31 LOWN, B., VERRIER, R.L.: Neural activity and ventricular fibrillation. New Engl. J. Med. *294*, 1165 (1976)
32 LOWN, B., WOLF, M.: Approaches to sudden death from coronary heart disease. Circulation *44*, 130 (1971)
33 MAROKO, P.R., KJEKSHUS, J.K., SOBEL, B.E., WATANABE, T., COVELL, J.W., ROSS, J., Jr., BRAUNWALD, E.: Factors influencing infarct size following experimental coronary artery occlusions. Circulation *43*, 67 (1971)
34 MJØS, O.D.: Effect of free fatty acids on myocardial function and oxygen consumption in intact dogs. J. clin. Invest. *50*, 1386 (1971)
35 MURNAGHAN, M.F.: The effect of sympathomimetic amines on the ventricular fibrillation threshold in the rabbit isolated heart. Brit. J. Pharmacol. *53*, 3 (1975)
36 OLIVER, M.F.: Metabolic response during impending myocardial infarction. II. Clinical implications. Circulation *45*, 491 (1972)
37 OLIVER, M.F., KURIEN, V.A., GREENWOOD, T.W.: Relation between serum-free-fatty-acids and arrhythmias and death after acute myocardial infarction. Lancet *1968/I*, 710
38 PAPP, J.G., VAUGHAN WILLIAMS, E.M.: A comparison of the anti-arrhythmic actions of I.C.I. 50172 and (−)-propranolol and their effects on intracellular cardiac action potentials and other features of cardiac function. Brit. J. Pharmacol. *37*, 391 (1969)
39 PENTECOST, B.L., AUSTEN, W.G.: Beta-adrenergic blockade in experimental myocardial infarction. Amer. Heart J. *72*, 790 (1966)
40 PITT, B., CRAVEN, P.: Effect of propranolol on regional myocardial blood flow in acute ischaemia. Cardiovasc. Res. *4*, 176 (1970)
41 PURI, P.S.: Modification of experimental myocardial infarct size by cardiac drugs. Amer. J. Cardiol. *33*, 521 (1974)
42 REDWOOD, D.R., SMITH, E.R., EPSTEIN, S.E.: Coronary artery occlusion in the conscious dog. Effects of alterations in heart rate and arterial pressure on the degree of myocardial ischemia. Circulation *46*, 323 (1972)
43 REIMER, K.A., RASMUSSEN, M.M., JENNINGS, R.B.: On the nature of protection by propranolol against myocardial necrosis after temporary coronary occlusion in dogs. Amer. J. Cardiol. *37*, 520 (1976)

44 Renaud, S., Kuba, K., Goulet, C., Lemire, Y., Allard, C.: Relationship between fatty-acid composition of platelets and platelet aggregation in rat and man. Relation to thrombosis. Circulat. Res. *26,* 553 (1970)
45 Rubegni, M., Provvedi, D., Bellini, P.G., Bandinelli, C., De Mauro, G.: Propranolol and platelet aggregation. Circulation *52,* 964 (1975); corresp.
46 Schaal, S.F., Wallace, A.G., Sealy, W.C.: Protective influence of cardiac denervation against arrhythmias of myocardial infarction. Cardiovasc. Res. *3,* 241 (1969)
47 Shell, W.E., Sobel, B.E.: Deleterious effects of increased heart rate on infarct size in the conscious dog. Amer. J. Cardiol. *31,* 474 (1973)
48 Shell, W.E., Sobel, B.E.: Protection of jeopardized ischemic myocardium by reduction of ventricular afterload. New Engl. J. Med. *291,* 481 (1974)
49 Sommers, H.M., Jennings, R.B.: Ventricular fibrillation and myocardial necrosis after transient ischemia. Arch. intern. Med. *129,* 780 (1972)
50 Sweet, C.S., Gaul, S.L.: Attenuation of hydrochlorothiazide-induced hypokalemia in dogs by a β-adrenergic-blocking drug, timolol. Europ. J. Pharmacol. *32,* 370 (1975)
51 Theroux, P., Franklin, D., Ross, J., Jr., Kemper, W.S.: Regional myocardial function during acute coronary artery occlusion and its modification by pharmacologic agents in the dog. Circulat. Res. *35,* 896 (1974)
52 Trautwein, W., Gottstein, U., Dudel, J.: Der Aktionsstrom der Myokardfaser im Sauerstoffmangel. Pflügers Arch. *260,* 40 (1954)
53 Verrier, R.L., Thompson, P.L., Lown, B.: Ventricular vulnerability during sympathetic stimulation: role of heart rate and blood pressure. Cardiovasc. Res. *8,* 602 (1974)
54 Wallace, A.G., Klein, R.F.: Role of catecholamines in acute myocardial infarction. Amer. J. med. Sci. *258,* 139 (1969)
55 Watanabe, T., Shintani, F., Fu, L., Fujii, J., Watanabe, H., Kato, K.: Influence of inotropic alteration on the severity of myocardial ischemia after experimental coronary occlusion. Jap. Heart J. *13,* 222 (1972)
56 Webb, S.W., Adgey, A.A.J., Pantridge, J.F.: Autonomic disturbance at onset of acute myocardial infarction. Brit. med. J. *1972/III,* 89
57 Wiggers, C.J., Wégria, R., Piñera, B.: The effects of myocardial ischemia on the fibrillation threshold – the mechanism of spontaneous ventricular fibrillation following coronary occlusion. Amer. J Physiol. *131,* 309 (1940)
58 Wit, A.L., Hoffman, B.F., Rosen, M.R.: Electrophysiology and pharmacology of cardiac arrhythmias. IX. Cardiac electrophysiologic effects of beta adrenergic receptor stimulation and blockade. Part C. Amer. Heart J. *90,* 795 (1975)
59 Wollenberger, A., Shahab, L.: Anoxia-induced release of noradrenaline from the isolated perfused heart. Nature (Lond.) *207,* 88 (1965)

Summary of discussion

Prof. GROSS opened the discussion on the paper presented by Prof. BRUNNER.

Dr. MORET requested further clarification of the manner in which propranolol influences the release of adrenaline and noradrenaline in the myocardium.

Prof. BRUNNER replied by explaining first of all that, in experimentally induced myocardial infarction, the local release of catecholamines was a consequence of myocardial hypoxia and that this hypoxia also affected the sympathetic nerve endings, which are rich in catecholamines. The release of catecholamines was a typical occurrence in those zones of the heart muscle that become hypoxic. Prior treatment with a beta-blocker served to diminish the degree of hypoxia, and this in turn measurably reduced the release of catecholamines in the affected area.

Dr. RANGOONWALA expressed regret that, in connection with the possible cardioprotective action of the beta-blockers, Prof. BRUNNER had not referred in his paper to their effects on the central nervous system, including especially their stress-attenuating properties.

Prof. GROSS suggested that this question be discussed later, since it was a problem upon which papers to be presented by other speakers would no doubt touch.

Prof. KÜBLER said that he could not quite agree with Prof. BRUNNER's interpretation of the release of catecholamines in the damaged myocardium. Their release was surely an immediate reaction, which occurred so quickly that it could hardly be accounted for in terms of the severity of the tissue damage sustained. Moreover, quantitative data on the release of catecholamines could not be regarded as fully reliable if they were based simply upon measurement of the catecholamine concentration in coronary sinus blood; to obtain really reliable figures, one would also have to measure the coronary blood flow, particularly after a drug had been administered which exerted an influence on coronary blood flow.

Prof. BRUNNER concurred with the latter observation and added that in his paper he had merely summarised the findings published on this topic. He also emphasised that it was not only the sympathetic nerve endings which participated in the release of catecholamines.

Dr. TAYLOR expressed interest in the question of dose-response relationships and dose equivalents in connection with the experimental findings which had been mentioned by Prof. BRUNNER.

Prof. BRUNNER pointed out in reply that the doses of beta-blockers employed in the studies on experimental infarction and experimental myocardial hypoxia were approximately equivalent to those used in patients suffering from angina pectoris.

Dr. SCHLESINGER requested further information on coronary blood flow in beta-blocked animals with experimentally induced myocardial hypoxia.

Prof. BRUNNER stressed that in response to beta-blockade both myocardial oxygen consumption and coronary blood flow were reduced to the same extent. This, however, applied only to that portion of the heart muscle which was adequately supplied with oxygen. The hypoxic area, by contrast, was evidently subject to a strong local vasodilator stimulus, with the result that under beta-blockade the blood flow in this area was relatively increased.

Beta-blockers and cardiac involvement in hypertension

by A. ZANCHETTI*

There are two aspects in which the heart may be considered as being involved in arterial hypertension, and, depending upon which of these two viewpoints one adopts, one can regard the beta-blockers either as tools for understanding the mechanisms of hypertension or as means for interfering with its development and consequences. Viewed from the first aspect, cardiac involvement implies that the heart plays a pathogenetic role, i.e. that it is primarily implicated in causing or mediating arterial hypertension; in this context, beta-blockers can be looked upon as tools with which to clarify this pathogenetic role and assess its importance or, alternatively, as therapeutic agents affording protection to the heart insofar as they specifically prevent the development of hypertension. Viewed from the second aspect, cardiac involvement can be regarded as a secondary phenomenon, in which the heart constitutes one of the targets, though a very important one, of the damaging action of high blood pressure. Accordingly, the beta-blockers can be seen here either as non-specific cardioprotective drugs – like all the other blood-pressure lowering substances – or as specific cardioprotective drugs which serve to protect the heart independently of their antihypertensive effect.

1. Primary or causative involvement of the heart in hypertension

One hypothesis which has been the subject of lively debate in recent years postulates that an increase in cardiac output may be the primary haemodynamic determinant of human and experimental hypertension. According to the concept of whole-body autoregulation so effectively marshalled by GUYTON and COLEMAN[17], a primary increase in cardiac output would later trigger an autoregulatory increase in peripheral resistance. Although, from the physiological standpoint, whole-body autoregulation is a mechanism which still requires to be substantiated and clarified, it is commonly accepted that in a considerable proportion of patients with borderline hypertension cardiac output is indeed elevated[8, 20, 33], whereas in established hypertension cardiac output is normal and peripheral resistance raised[11].

There are various mechanisms by which cardiac output may be initially raised in hypertension, and not all of these mechanisms are due to a primary cardiac disturbance. The one first suggested in 1963 by BORST and BORST-DE GEUS[3], and later by LEDINGHAM[22] and GUYTON et al.[18], is a primary disorder of renal origin in which sodium retention leads to expansion of plasma volume and consequently to an increased cardiac output.

Another possible extracardiac mechanism is an increase in central blood volume resulting in increased filling of the heart[24], but there is no general agreement on this

---

* Istituto di Ricerche Cardiovascolari e Istituto di Patologia Medica I dell'Università di Milano, Milan, Italy.

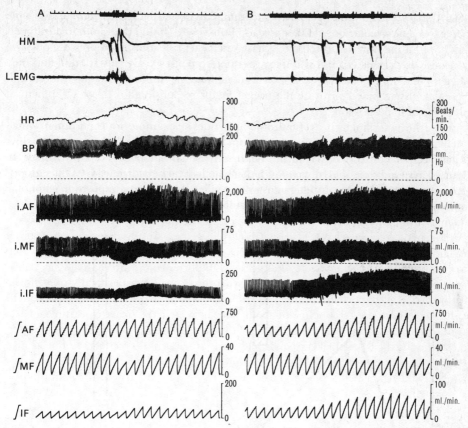

Fig. 1. Cardiovascular changes occurring in a free-moving cat during a brief (A) and a prolonged (B) period of fighting.
Shown here (reading from top to bottom) are: time, graduated in units of 1 and 5 seconds, and periods of fighting (indicated by the thick bars); HM = head movements; L.EMG = electromyogram of the left forelimb (signalling striking movements); HR = heart rate; BP = arterial blood pressure; i.AF = instantaneous aortic flow (cardiac output); i.MF = instantaneous superior mesenteric flow; i.IF = instantaneous left external iliac flow; ∫AF, ∫MF, and ∫IF = 2-second integrations of the above flows. (From: ADAMS et al.[1], unpublished figure)

issue. ELLIS and JULIUS[9], as well as TARAZI et al.[28], point out, for example, that the heart of the borderline hypertensive patient responds to normal venous filling with an increased stroke volume. According to JULIUS and ESLER[20], this enhanced response is neurogenic.

Neural mechanisms are known to affect the heart, and in this connection emotional influences have been studied in particular detail. Illustrated in Figure 1 is an example from our own studies on the haemodynamic manifestations of fighting behaviour in cats[1, 34]. It shows the marked increase in heart rate and cardiac output occurring during both short and more prolonged episodes of fighting: the increase in arterial pressure is largely maintained via this augmented cardiac output, whereas total peripheral resistance is decreased.

As reported by BROD[4], a similar haemodynamic pattern can be observed in normotensive and hypertensive subjects in response to the emotional stimulus of difficult mental arithmetic. Recent studies on spontaneously hypertensive rats of the Okamoto strain have revealed that, like borderline human hypertensives, these animals too have a hyperkinetic circulation[12]. HALLBACK and FOLKOW[10,19] found that, both in adulthood and even in the "pre-hypertensive" phase, the heart rate and blood pressure of spontaneously hypertensive rats exhibit exaggerated responses to alerting stimuli, and they suggest that more frequent and powerful neurogenic stimuli reaching the cardiovascular system may constitute the trigger factor responsible for the development of this type of hypertension.

Finally, the hyperkinetic heart syndrome[14] provides an example of a condition in which an increase in cardiac output may result from a primary disturbance of the heart – since it has been demonstrated that in this syndrome the cardiac beta-receptors display enhanced responsiveness[13]. In cases of hyperkinetic heart syndrome, GUAZZI's group[2] have shown that the hyperkinetic signs worsen when the patient is exposed to emotional stimuli such as the performance of a difficult arithmetic task.

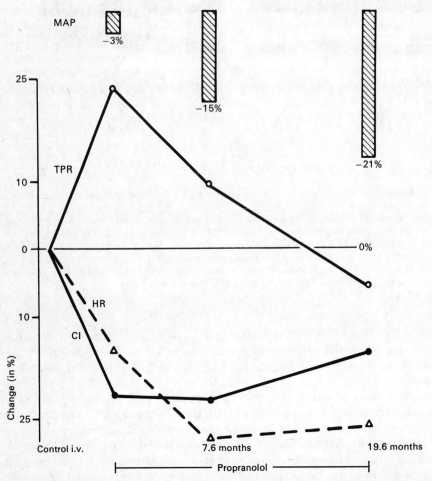

Fig. 2. Changes recorded in ten hypertensive patients in response to intravenous injection of propranolol and to long-term oral treatment with this beta-blocker. The changes in cardiac index (CI), total peripheral resistance (TPR), and heart rate (HR), as well as the decrease in mean arterial pressure (MAP), are all expressed in percent of the pre-treatment values. (From: TARAZI and DUSTAN[27], by courtesy of the American Journal of Cardiology)

How can the beta-blockers help us in clarifying the question whether the heart plays a primary role in the origin of hypertension and in unravelling the various pathogenetic mechanisms in which it may be involved?

Studies undertaken during recent years by TARAZI and DUSTAN [27] have yielded findings which appear to provide consistent support for the notion of primary cardiac involvement in the pathogenesis of hypertension. One of the experiments performed by these authors is outlined in Figure 2. Here it can be seen that intravenous injection of a beta-blocker, though causing an immediate and conspicuous reduction in cardiac output, does not result in lowering of the blood pressure, because the decrease in cardiac output is offset by an increase in peripheral resistance. It is only in response to chronic oral administration that the beta-blocker exerts an antihypertensive effect, the reduction in cardiac output now being accompanied by no change, or even a decrease, in peripheral resistance. The conclusion drawn from this interesting experiment is that, in patients receiving prolonged treatment with a beta-blocker, the blood pressure decreases owing to progressive autoregulatory vasodilatation occurring in response to the chronic fall in cardiac output. An obvious corollary is that hypertension may develop as a result of progressive autoregulatory vasoconstriction provoked by a prolonged increase in cardiac output.

Now, however, it is possible to place an alternative interpretation on the observations made by TARAZI and DUSTAN. Recent findings obtained by our group [23] indicate that both the bradycardic and the antihypertensive effects of propranolol are dose-dependent, but that the dose-response curves for these two effects are different. As shown in Figure 3, propranolol already exerts quite a marked effect on heart rate at very low plasma concentrations, i.e. concentrations which are devoid of any antihypertensive

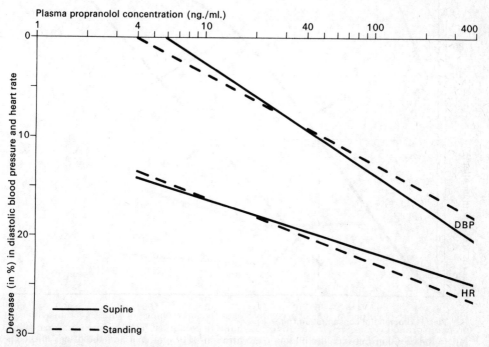

Fig. 3. Relationship of changes in diastolic blood pressure (DBP) and in heart rate (HR) to the logarithm of the plasma propranolol concentration. (From: LEONETTI et al.[23], unpublished figure)

action. No decrease in blood pressure occurs until the plasma concentrations of propranolol reach levels at which no further bradycardia is observed. Even during strenuous exercise, inhibition of exertional tachycardia attains its maximum at plasma propranolol concentrations of about 100 ng./ml. This observation has enabled us to calculate an ED_{50} for the effect of propranolol on heart rate – in other words, the plasma concentration at which this beta-blocker exerts 50% of its maximum bradycardic activity[6]. As indicated in Figure 4, this ED_{50} is approximately 8 ng./ml. It is impossible to calculate an ED_{50} for the drug's blood-pressure lowering effect, because even at the highest plasma concentrations recorded in our patients the curve had not yet attained a plateau. But, even if the decrease in blood pressure observed at 400 ng./ml. were to be regarded as equivalent to the drug's maximum effect, the corresponding ED_{50} would not be less than 40 ng./ml.; it should be made clear, in fact, that the interrupted diagonal line in Figure 4 simply represents a borderline, to the right of which the real curve would probably lie. At all events, the wide difference between the curves for the bradycardic and the antihypertensive effects of propranolol does not support the concept that the fall in blood pressure is necessarily due to autoregulatory mechanisms called into operation by a chronic reduction in cardiac output.

There are other data available which also seem to cast doubt on the notion that the heart plays a primary role in hyperkinetic types of hypertension, such as human borderline hypertension and spontaneous hypertension in Okamoto rats. JULIUS et al.[21] have reported that, after total autonomic cardiac blockade with propranolol and atropine has markedly reduced the elevated cardiac index of hyperkinetic borderline hyper-

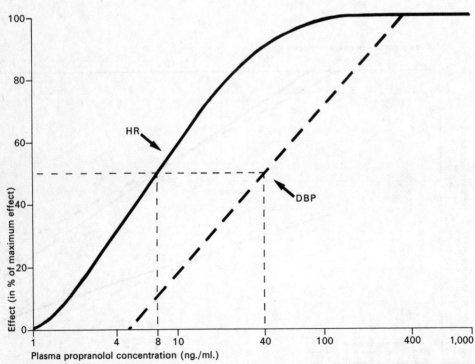

Fig. 4. Relationship between the plasma concentration of propranolol and the drug's effect on heart rate (HR) and diastolic blood pressure (DBP), expressed in percent of the maximum effect (for further details, see text).

tensives, their blood pressure still remains elevated owing to a rise in the previously normal or low peripheral vascular resistance. FROHLICH and PFEFFER[12] have likewise shown that in spontaneously hypertensive rats continuous beta-adrenergic blockade by sotalol from the fourth to the seventh week of age does not prevent the progressive development of hypertension, despite a marked reduction in heart rate. This suggests that the increased cardiac output of these animals is not an essential prerequisite for the development of hypertension. It is true that WEISS et al.[32] have been able to prevent further progression of hypertension in spontaneously hypertensive rats by administering large doses of propranolol or of metoprolol (®Lopresor), but it would appear likely that – in accordance with the concept I have already outlined by reference to the differing dose-response curves illustrated in Figure 4 – the very large doses employed by these authors would also have produced an antihypertensive effect independently of the effect on the heart.

The use of beta-blocking drugs both in human and in experimental hypertension has thus failed to provide any crucial support for the theory that the heart is primarily involved in causing or maintaining arterial hypertension. Although this is still an open question, there are no grounds at present for claiming that beta-blockers might exert a cardioprotective action by specifically influencing a cardiac mechanism which leads to the development of hypertension.

This does not exclude the possibility, however, that beta-blockers may exert a cardioprotective action in hypertension by other means.

2. Secondary involvement of the heart in hypertension

It is a well-known fact that chronic hypertension is liable to result in damage to the heart, and several epidemiological studies have shown that coronary disease becomes increasingly frequent the higher the arterial pressure[7]. This being the case, it was rather surprising that the results obtained by the Veterans Administration Cooperative Study Group on Antihypertensive Agents[31], though clearly demonstrating the effectiveness of antihypertensive therapy in preventing renal and cerebral vascular damage, failed to show any significant improvement in response to such treatment so far as coronary artery disease was concerned. It therefore seems reasonable to speculate whether a different therapeutic regimen might perhaps have achieved a better cardioprotective effect. The treatment employed in the Veterans Administration Study generally included a diuretic, reserpine, and hydralazine. There are at least two reasons why antihypertensive treatment including a beta-blocker might be expected to afford a greater degree of protection for the heart. The first is the argument that renin represents, as it were, a risk factor in the development of vascular damage secondary to hypertension[5], and that renin suppression, which is so easily produced by beta-blockade[36], might thus help to provide improved protection. However, quite apart from the fact that such protection should preferably not be specifically confined to the heart but should extend to all the vascular targets of hypertension, conclusive proof of the role played by renin as a risk factor is still lacking and may possibly never be forthcoming. A more likely explanation for the specific cardioprotective action of the beta-blockers is to be found in the antagonistic effect which they exert on sympathetic activation of the heart. Assuming that repeated sympathetic excitation under stressful conditions is of importance in promoting coronary disease or – as seems more likely – in precipitating a coronary attack[35], this factor might well prove even more detrimental in hearts already impaired by hypertension. Figure 5, reproduced from a paper by TAYLOR[29], shows that oxpren-

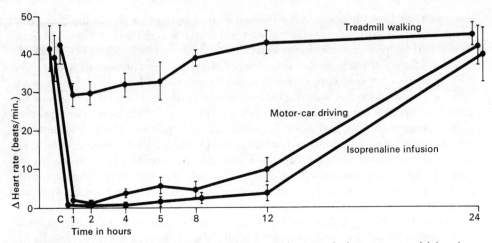

Fig. 5. Effect of 40 mg. oral oxprenolol on the heart rate increase during motor-car driving, isoprenaline infusion, and treadmill walking in normal subjects. The control values (C) are the average of the increases in heart rate measured throughout the day on placebo. Data expressed as mean of observations ± S.E.M. (From: TAYLOR[29])

olol (®Trasicor) is highly effective in preventing emotional tachycardia such as that induced by motor-car driving, and that it already achieves this effect when given in small doses which only partially counteract exertional tachycardia. There are thus theoretical grounds for postulating as a working hypothesis that beta-blockers may be potentially capable of protecting the hearts of hypertensive patients.

The results of a prospective study recently reported by STEWART[26] suggest that such a protective action might indeed exist. In this study hypertensive patients, matched for age and diastolic blood pressure, were followed up for a mean period of 5.25 years while undergoing antihypertensive therapy. Of 121 patients whose regimen included a beta-blocker, prescribed either alone or in combination with other drugs, only nine (7.5%) suffered a myocardial infarction during treatment, as compared with no fewer than 15 (31.2%) out of 48 patients whose blood pressure was equally well controlled by an antihypertensive regimen not including a beta-blocker. These data are certainly impressive, and they would appear to indicate that beta-blockers might well exert a specific cardioprotective action, which deserves to be more carefully and more extensively investigated – particularly since the study undertaken by STEWART, like most other pioneering studies, is open to certain criticisms. Firstly, the incidence of myocardial infarction in the patients not receiving a beta-blocker (31.2%) is very high for a treated group – considerably higher, in fact, than in patients with comparable diastolic pressures participating in the Veterans Administration Study. Secondly, the effect of the various antihypertensive regimens on heart rate should have been assessed comparatively, especially since a number of the drugs administered to the group of 48 patients, such as clonidine, guanethidine, and bethanidine, also have the effect of reducing the heart rate. Finally, since beta-blockers exert their action on the heart in doses lower than those required to produce an antihypertensive effect, it is reasonable to suppose that the addition of a small dose of a beta-blocker to a conventional antihypertensive regimen might already suffice to improve cardiac prognosis without necessarily reinforcing the antihypertensive response; this possibility is of the utmost interest and would be well worth investigating.

Although the cardioprotective action of the beta-blockers in hypertension is a definite possibility which simply requires conclusive confirmation, it should be borne in mind that the risk of a detrimental effect also deserves attention. When considering the action of the sympathetic system on the heart, it is always difficult to distinguish the deleterious consequences of overwork from the supportive functions implicit in the well-known shift to the left in Starling's volume-pressure curve[25]. In a paper published a few years ago, GUAZZI et al.[15] stressed the importance of sympathetic drive in helping to prevent heart failure in the hypertensive patient. The studies upon which this conclusion was based have meanwhile been extended by GUAZZI's group[16] to include investigations on the influence exerted by beta-blockers on cardiac performance in hypertensive patients. It is interesting to note that these authors have observed three different patterns of haemodynamic responses to prolonged oral administration of propranolol, i.e. to daily doses of 320 mg. given for three weeks. The patients concerned were divided into responders and non-responders, depending on whether or not the mean blood pressure decreased under treatment with the beta-blocker. In the non-responders, cardiac index was reduced and total peripheral resistance increased, i.e. impedance to ejection was augmented and the mean systolic ejection rate consequently decreased; a poorer left-ventricular performance in the pre-ejection period was also reflected in a decrease in the mean pre-ejection dp/dt. Among the responders, it was possible to differentiate between two groups. In one group the fall in blood pressure was associated with a decrease in cardiac output and an increase in peripheral resistance; despite the fall in mean blood pressure, impedance to ejection was augmented as in the non-responders, and the mean systolic ejection rate therefore decreased, indicating a poorer ejection performance on the part of the left ventricle. Pre-ejection performance was also reduced. In the other group of responders, by contrast, the fall in blood pressure induced by propranolol was accompanied by little change in cardiac output and by a decrease in peripheral resistance; in these patients the improvement in left-ventricular ejection was reflected in a rise, rather than a diminution, in the mean systolic ejection rate.

It would appear from these interesting findings that in hypertensive patients a fall in the elevated blood pressure levels does not always go hand in hand with a decrease in the afterload and an improvement in the ejection performance of the left ventricle. On the contrary, this is true only of a certain proportion of patients; and, while not all hypertensives displaying a reduced ejection performance will necessarily develop heart failure, it is among these hypertensives that cases of drug-induced heart failure are likely to occur. Unfortunately, no criteria are yet available for predicting which haemodynamic pattern a given patient will exhibit. Whether beta-blockers such as oxprenolol or pindolol, which possess an intrinsic sympathomimetic action and which thus tend to stimulate cardiac function, might have a certain additional advantage to offer, is not known at present, although this is a point that certainly merits investigation.

Conclusions

To sum up, it may be said that beta-blockers are definitely endowed with antihypertensive properties and that for this reason they can be regarded as cardioprotective agents at least to the same extent as all other drugs displaying comparable antihypertensive efficacy. At the moment, however, it would not seem that the beta-blockers have a specific antihypertensive action, i.e. an action affecting the cause, or a causative mechanism, of hypertension. On the other hand, that the beta-blockers may be capable of playing a cardioprotective role in a way which other antihypertensive agents cannot, is a postulate that can now be envisaged insofar as blockade of the beta-receptors might serve to protect the hypertensive heart from the additional strain of excessive sympathetic activity. Although this is still more in the nature of an hypothesis than an established fact, preliminary data suggest that beta-blockers might improve the prognosis for the hypertensive patient to a greater extent than do other drugs such as those employed in the Veterans Administration Study. But, if full advantage is to be taken of the clinical benefits resulting from the beta-blockers in hypertension, physicians must be aware of the possibility that – at least in a few patients – these drugs may produce detrimental effects, and research workers must strive to identify those clues which might distinguish the minority of hypertensive patients in whom beta-blockers are not unreservedly indicated from the majority who certainly would benefit from their use.

References

1. ADAMS, D.B., BACCELLI, G., MANCIA, G., ZANCHETTI, A.: Cardiovascular changes during naturally elicited fighting behavior in the cat. Amer. J. Physiol. *216*, 1226 (1969)
2. BARTORELLI, C., POLESE, A., FIORENTINI, C., MAGRINI, F., OLIVARI, M.T., GUAZZI, M.: Electrical and dynamic responses of the human hyperkinetic heart to sympathetic stimuli. Clin. Sci. molec. Med. *48*, Suppl. 2:291 (1975)
3. BORST, J.G.G., BORST-DE GEUS, A.: Hypertension explained by Starling's theory of circulatory homoeostasis. Lancet *1963/I*, 677
4. BROD, J.: Haemodynamic basis of acute pressor reactions and hypertension. Brit. Heart J. *25*, 227 (1963)
5. BRUNNER, H.R., LARAGH, J.H., BAER, L., NEWTON, M.A., GOODWIN, F.T., KRAKOFF, L.R., BARD, R.H., BÜHLER, F.R.: Essential hypertension: renin and aldosterone, heart attack and stroke. New Engl. J. Med. *286*, 441 (1972)
6. CHIDSEY, C.A., ZANCHETTI, A., MORSELLI, P., LEONETTI, G.: Pharmacokinetic and pharmacodynamic studies of propranolol in hypertension. In Milliez, P., Safar, M. (Editors): Recent advances in hypertension, p. 319 (Reims 1975)
7. DAWBER, T.R., KANNEL, W.B., REVOTSKIE, N., KAGAN, A.: The epidemiology of coronary heart disease – the Framingham enquiry. Proc. roy. Soc. Med. *55*, 265 (1962)
8. EICH, R.H., PETERS, R.J., CUDDY, R.P., SMULYAN, H., LYONS, R.H.: The hemodynamics in labile hypertension. Amer. Heart J. *63*, 188 (1962)
9. ELLIS, C.N., JULIUS, S.: Role of central blood volume in hyperkinetic borderline hypertension. Brit. Heart J. *35*, 450 (1973)
10. FOLKOW, B.: Central neurohormonal mechanisms in spontaneously hypertensive rats compared with human essential hypertension. Clin. Sci. molec. Med. *48*, Suppl. 2:205 (1975)
11. FREIS, E.D.: Hemodynamics of hypertension. Physiol. Rev. *40*, 27 (1960)
12. FROHLICH, E.D., PFEFFER, M.A.: Adrenergic mechanisms in human hypertension and in spontaneously hypertensive rats. Clin. Sci. molec. Med. *48*, Suppl. 2:255 (1975)
13. FROHLICH, E.D., TARAZI, R.C., DUSTAN, H.P.: Hyperdynamic β-adrenergic circulatory state. Increased β-receptor responsiveness. Arch. intern. Med. *123*, 1 (1969)
14. GORLIN, R.: The hyperkinetic heart syndrome. J. Amer. med. Ass. *182*, 823 (1962)
15. GUAZZI, M., MAGRINI, F., FIORENTINI, C., POLESE, A.: Role of the sympathetic nervous system in supporting cardiac function in essential hypertension. Brit. Heart J. *35*, 55 (1973)
16. GUAZZI, M., POLESE, A., FIORENTINI, C., OLIVARI, M.T., MAGRINI, F., BARTORELLI, C.: Cardiac function in the treatment of arterial hypertension with propranolol. Clin. Sci. molec. Med. *51*, Suppl. 3: 555s (1976)
17. GUYTON, A.C., COLEMAN, T.G.: Quantitative analysis of the pathophysiology of hypertension. Circulat. Res. *24*, Suppl. I:1 (1969)
18. GUYTON, A.C., COLEMAN, T.G., COWLEY, A.W., Jr., SCHEEL, K.W., MANNING, R.D., NORMAN, R.A., Jr.: Arterial pressure regulation. Overriding dominance of the kidneys in long-term regulation and in hypertension. Amer. J. Med. *52*, 584 (1972)

19 HALLBACK, M., FOLKOW, B.: Cardiovascular responses to acute mental "stress" in spontaneously hypertensive rats. Acta physiol. scand. *90*, 684 (1974)
20 JULIUS, S., ESLER, M.: Autonomic nervous cardiovascular regulation in borderline hypertension. Amer. J. Cardiol. *36*, 685 (1975)
21 JULIUS, S., ESLER, M. D., RANDALL, O. S.: Role of the autonomic nervous system in mild human hypertension. Clin. Sci. molec. Med. *48,* Suppl. 2:243 (1975)
22 LEDINGHAM, J. M.: The etiology of hypertension. Practitioner *207*, 5 (1971)
23 LEONETTI, G., MAYER, G., MORGANTI, A., TERZOLI, L., ZANCHETTI, A., BIANCHETTI, G., DI SALLE, E., MORSELLI, P. L., CHIDSEY, C. A.: Hypotensive and renin-suppressive activities of propranolol in hypertensive patients. Clin. Sci. molec. Med. *48*, 491 (1975)
24 SAFAR, M. E., WEISS, Y. A., LONDON, G. M., FRANCKOWIAK, R. F., MILLIEZ, P. L.: Cardiopulmonary blood volume in borderline hypertension. Clin. Sci. molec. Med. *47*, 153 (1974)
25 SARNOFF, S. J., MITCHELL, J. H.: The control of the function of the heart. In: Handbook of Physiology, Circulation. Vol. 1, Chap. 15, p. 489 (American Physiological Association, Washington 1962)
26 STEWART, I. McD. G.: Compared incidence of first myocardial infarction in hypertensive patients under treatment containing propranolol or excluding β-receptor blockade. Clin. Sci. molec. Med. *51*, Suppl. 3:509s (1976)
27 TARAZI, R. C., DUSTAN, H. P.: Beta-adrenergic blockade in hypertension. Practical and theoretical implications of long-term hemodynamic variations. Amer. J. Cardiol. *29*, 633 (1972)
28 TARAZI, R. C., IBRAHIM, M. M., DUSTAN, H. P., FERRARIO, C. M.: Cardiac factors in hypertension. Circulat. Res. *34/35*, Suppl. 1:213 (1974)
29 TAYLOR, S. H.: The role of stress factors in ischaemic heart disease and their modulation by beta-receptor antagonists. In Schweizer, W. (Editor): Beta-blockers – present status and future prospects. Int. Symp., Juan-les-Pins 1974, p. 167 (Huber, Berne/Stuttgart/Vienna 1974)
30 VETERANS ADMINISTRATION COOPERATIVE STUDY GROUP ON ANTIHYPERTENSIVE AGENTS: Effects of treatment on morbidity in hypertension. Results in patients with diastolic blood pressures averaging 115 through 129 mm Hg. J. Amer. med. Ass. *202*, 1028 (1967)
31 VETERANS ADMINISTRATION COOPERATIVE STUDY GROUP ON ANTIHYPERTENSIVE AGENTS: Effects of treatment on morbidity in hypertension. II. Results in patients with diastolic blood pressure averaging 90 through 114 mm Hg. J. Amer. med. Ass. *213*, 1143 (1970)
32 WEISS, L., LUNDGREN, Y., FOLKOW, B.: Effects of prolonged treatment with adrenergic β-receptor antagonists on blood pressure, cardiovascular design and reactivity in spontaneously hypertensive rats (SHR). Acta physiol. scand. *91*, 447 (1974)
33 WIDIMSKY, J., FEJFAROVA, M. H., FEJFAR, Z.: Changes of cardiac output in hypertensive disease. Cardiologia *31*, 381 (1957)
34 ZANCHETTI, A., BACCELLI, G., MANCIA, G., ELLISON, G. D.: Emotion and the cardiovascular system in the cat. In: Physiology, emotion and psychosomatic illness, Ciba Foundation Symposium 8, p. 201 (new series). (Associated Scientific Publishers, Amsterdam 1972)
35 ZANCHETTI, A., MALLIANI, A.: Neural and psychological factors in coronary disease. Acta cardiol., Suppl. XX:69 (1974)
36 ZANCHETTI, A., STELLA, A., LEONETTI, G., MORGANTI, A., TERZOLI, L.: Control of renin release: a review of experimental evidence and clinical implications. Amer. J. Cardiol. *37*, 675 (1976)

Effects of beta-blockade on diurnal variability of blood pressure

by W. H. Birkenhäger, P. W. de Leeuw, H. E. Falke, and A. Wester *

Diurnal fluctuations in blood pressure are probably relevant to the problem of "stabilising" myocardial function in the hypertensive patient.

The 24-hour blood pressure patterns, and the influence exerted upon them by sleep in particular, have been extensively studied by various authors [2-4, 6-9] both in normotensives and in hypertensives. For this purpose, direct as well as indirect automatic techniques have been used. To our knowledge, however, no reports have yet been published on the effects of beta-adrenergic blocking agents in this connection, although Clement[5], working in Ghent, has just completed a study which is rather similar to ours.

The investigations to be described here were designed to assess the influence of a beta-blocker (propranolol) on this aspect of blood pressure control.

Sixteen subjects with uncomplicated essential hypertension (mean arterial pressure: 110–130 mm. Hg) were admitted to hospital, where their sodium intake was fixed at 60 mmol. per day and monitored by collecting and analysing the 24-hour urine. Their blood pressure was registered day and night at 20-minute intervals, using an automatic recording system involving indirect measurement of the blood pressure ("Arteriosonde"/Roche). It was felt that the use of an invasive technique would interfere with natural sleep patterns, whereas acclimatisation to the indirect recording procedure was rapidly achieved after a run-in period of one or two nights. For computation of the blood pressure readings we developed the following procedure (illustrated in Figure 1):

The "basal" blood pressure measured immediately after waking (in accordance with the criterion adopted by Alam and Smirk[1]) was employed as a reference value. The difference between this basal value and the maximum upward fluctuation in blood

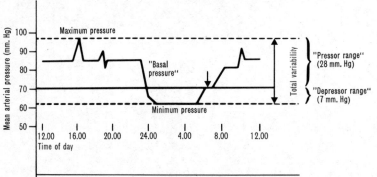

Fig. 1. Diagrammatic representation of diurnal blood pressure fluctuations and indices of blood pressure variability as used in the present study.

* Afdeling Inwendige Geneeskunde, Zuiderziekenhuis, Rotterdam, Netherlands.

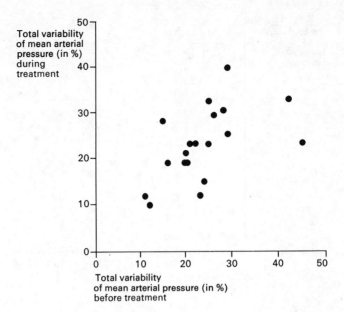

Fig. 2. Findings indicating that the "total variability" of the mean arterial pressure (expressed in percent of the "basal pressure") was not altered by beta-blockade (r = 0.53, P < 0.05).

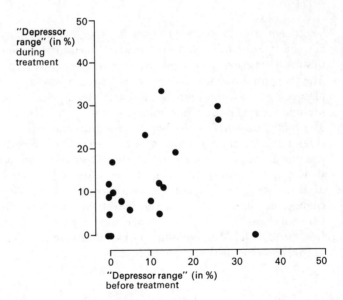

Fig. 3. Findings indicating that the "depressor range" (i.e. the nocturnal fall in blood pressure) appeared to be amplified during treatment (r = 0.36, not significant).

pressure recorded during the day was termed the "pressor range"; and the difference between the basal value and the lowest level of blood pressure recorded during sleep was designated the "depressor range". The sum of both ranges thus represented the "total variability" of the blood pressure. These values can be expressed either in mm. Hg or as a percentage of the basal blood pressure. Other variables measured were the fasting plasma renin and plasma noradrenaline concentrations, for the determination of which blood samples were taken at 10 a.m., i.e. while the patients were still in bed and had not yet breakfasted.

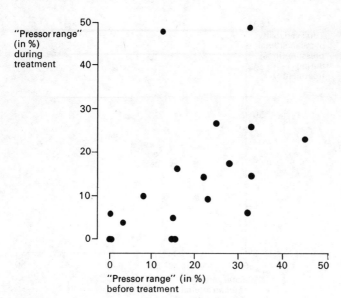

Fig. 4. Findings indicating that the "pressor range" (i.e. daytime rises in blood pressure from the basal level) tended, with a few exceptions, to decrease during treatment (r = 0.48, P < 0.05).

Results

Treatment with propranolol (average dose: 240 mg. daily) for ten days led to an average decrease in mean blood pressure (measured at noon) of 10%. At the end of this ten-day period the test procedure was repeated.

The individual response of the blood pressure was unrelated to the pre-treatment plasma renin concentration, but it did exhibit a (weak) positive correlation with the pre-treatment noradrenaline levels.

The total variability of the blood pressure remained unchanged during treatment (Figure 2). Within the range of total variability, however, a peculiar modulation was noted: the depressor range, i.e. the nocturnal fall in blood pressure during sleep, showed an increase in the majority of patients (Figure 3), and the pressor range decreased *pari passu* (Figure 4); this decrease in the pressor range was unrelated to changes in the plasma renin concentration. An inverse relationship was observed between changes in the pressor range and changes in the plasma noradrenaline levels (r = −0.59). The noradrenaline concentration rose in eight of the patients, diminished in five, and remained unchanged in three.

Discussion

The results reported here indicate that the overall effect of beta-blockade on blood pressure may be greater than would appear from standard evaluations. The increased nocturnal fall in blood pressure during sleep is of particular interest, because it suggests that some modulation may occur within the baroreflex loop. It remains to be seen whether this has favourable or unfavourable repercussions on myocardial perfusion.

On the other hand, the decrease in the pressor range recorded in these hospitalised patients during the daytime can only be regarded as a favourable response and is also consonant with observations to be reported later in this symposium by TAYLOR[10]. Surprisingly enough, the decrease in the pressor range showed an inverse relationship to the alterations occurring in the plasma noradrenaline levels. But how can this finding be reconciled with the general concept that acute increases in blood pressure are mediated through the adrenergic system? One explanation may be that the re-uptake of noradrenaline is affected by propranolol in such a way as to distort the relationships between adrenergic nerve traffic and net noradrenaline release. This would imply that noradrenaline determinations made during beta-blockade cannot be used as a parameter of adrenergic activity.

Finally, it should be emphasised that these observations of ours are valid only for periods of treatment lasting up to ten days. More prolonged treatment might conceivably induce a modification of this pattern.

Acknowledgment

The authors wish to thank Mr. R. PUNT for the technical assistance he provided in connection with this study, which was supported by Grant 732-61 from the Organisation for Health Research T.N.O., The Hague, Netherlands.

References

1 ALAM, G.M., SMIRK, F.H.: Casual and basal blood pressures. Brit. Heart J. *5*, 152 (1943)
2 ATHANASSIADIS, D., DRAPER, G.J., HONOUR, A.J., CRANSTON, W.I.: Variability of automatic blood pressure measurements over 24-hour periods. Clin. Sci. *36*, 147 (1969)
3 BEVAN, A.T., HONOUR, A.J., STOTT, F.H.: Direct arterial pressure recording in unrestricted man. Clin. Sci. *36*, 329 (1969)
4 BIRKENHÄGER, W.H., VAN ES, L.A., HOUWING, A., LAMERS, H.J., MULDER, A.H.: Studies on the lability of hypertension in man. Clin. Sci. *35*, 445 (1968)
5 CLEMENT, D.L.: (In preparation)
6 KHATRI, I.M., FREIS, E.D.: Hemodynamic changes during sleep. J. Appl. Physiol. *22*, 867 (1967)
7 RICHARDSON, D.W., HONOUR, A.J., FENTON, G.W., STOTT, F.H., PICKERING, G.W.: Variation in arterial pressure throughout the day and night. Clin. Sci. *26*, 445 (1964)
8 RICHARDSON, D.W., VETROVEC, G.W., WILLIAMSON, W.C.: In Onesti, G., Kim, K.E., Moyer, J.H. (Editors): Hypertension, mechanisms and management, XXVIth Hahnemann Symp., p. 141 (Grune & Stratton, New York/London 1973)
9 SHAW, D.B., KNAPP, M.S., DAVIES, D.H.: Variations of blood-pressure in hypertensives during sleep. Lancet *1963/I*, 797
10 TAYLOR, S.H.: Clinical aspects of the treatment of cardiovascular diseases with beta-blocking drugs. In Gross, F. (Editor): The cardioprotective action of beta-blockers, Int. Symp., Amsterdam 1976, p. 81 (Huber, Berne/Stuttgart/Vienna 1977)

The antihypertensive drug of first choice

by F. R. Bühler*

Among the various factors which promote cardiovascular disease in the general population, hypertension is evidently now the most common and most serious of the major contributors to morbidity and mortality. Coronary complications of hypertension are the leading cause of death, which in some 50% of cases takes the form of sudden death.

All the actuarial and epidemiological data thus far available, however, indicate that the use of drugs to reduce elevated blood pressure has not been associated with any decrease in the incidence of myocardial infarction, although it certainly has reduced the incidence of heart failure, renal insufficiency, and stroke, as well as the overall death rate among hypertensives[1,2]. Today it is a widely held belief that a diuretic drug should constitute the basic element of treatment for hypertension. In the Veterans Administration Study[1,2], for example, it was diuretics that were employed as the drugs of first choice.

Although the efficacy of diuretics as antihypertensive agents is beyond dispute, the question nevertheless arises as to whether these drugs, while lowering elevated blood pressure, might perhaps at the same time bring about other changes which could possibly enhance the risk of myocardial infarction. One of the effects which diuretics produce, for instance, is a rise in the plasma catecholamine levels – with all its attendant consequences. The influence exerted by diuretics on the adrenergic system also results in stimulation of the renin-angiotensin system, and both these systems are known to be capable of causing vasoconstriction and of promoting vascular damage. In addition, diuretics increase free fatty acid levels in the blood and decrease the plasma potassium concentration – two effects which are both liable to facilitate ventric-

Diuretics		Beta-blockers
↑	Catecholamines	Blockade of beta-receptors
↑	Angiotensin	↓
↑	Free fatty acids	↓
↓	Plasma potassium	↑
↓	Blood volume	~
↑	Platelet aggregation	↓
↑	Blood viscosity	↓

Fig. 1. Adrenergic, metabolic, and microcirculatory consequences of diuretic therapy and beta-blockade.

* Departement für Innere Medizin der Universität, Kantonsspital, Basle, Switzerland.

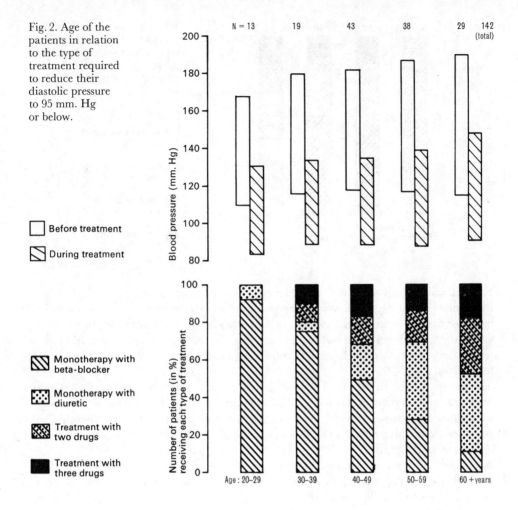

Fig. 2. Age of the patients in relation to the type of treatment required to reduce their diastolic pressure to 95 mm. Hg or below.

ular arrhythmias and to add to the risk of sudden death. What is more, in response to treatment with diuretics blood volume is reduced, the haematocrit value and the viscosity of the blood are increased, and platelet aggregation is stimulated; these changes are conducive to thrombo-embolic complications and also to vascular damage. On the other hand, as indicated in Figure 1, the use of beta-blockers entails none of these adrenergic, metabolic, and microcirculatory consequences. If such repercussions are indeed of significant clinical importance, then it might well be that treatment with beta-blockers could contribute to the prevention of myocardial infarction and sudden death. Moreover, by administering a beta-blocker in addition to a diuretic, it may be possible not only to potentiate the latter's antihypertensive activity but also to counteract the undesirable effects of diuretic therapy to which reference has just been made.

For these reasons – and also because the beta-blockers are generally acknowledged to be both therapeutically effective and relatively devoid of side effects – we carried out a trial in which treatment for hypertension was based primarily on the use of a beta-blocker. The patients studied, who consisted of hospitalised hypertensives with no clinical evidence of heart failure and no occurrences of bronchial asthma in their case histories, were initially given treatment with a beta-blocker only. If their diastolic

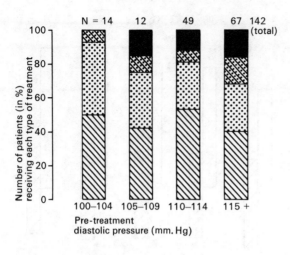

Fig. 3. Pre-treatment diastolic pressure in relation to the type of treatment required to reduce this pressure to 95 mm. Hg or below.

pressure failed to drop to the target level of 95 mm. Hg or below, a diuretic was then added to the regimen. Whenever possible, we later withdrew the beta-blocker in these cases, so as to test the efficacy of monotherapy with the diuretic. In instances where the target level could not be attained even in response to combined treatment with a beta-blocker and a diuretic, a third drug – i.e. a vasodilator – was prescribed in addition.

The results yielded by this triple-component treatment schedule were analysed after an average of 5.3 months. Of the total of 155 patients, 142 (92%) showed a decrease in diastolic pressure to 95 mm. Hg or below; the treatment had to be discontinued because of unwanted effects in eight cases, and because of inadequate blood pressure control in a further five.

A satisfactory response to monotherapy with the beta-blocker was obtained in 46% of the 142 patients; this response also showed a correlation with age, i.e. the beta-blocker reduced the diastolic pressure to or below the target level in 80% of the patients aged less than 40 years, in 30–50% of those aged 40–60 years, and in only 10% of those aged over 60 years (cf. Figure 2). Monotherapy with a diuretic lowered the diastolic pressure to or below 95 mm. Hg in another 25% of those patients who had not responded satisfactorily to the beta-blocker. In 15% of cases, the blood pressure was reduced to the desired level following treatment with the beta-blocker plus the diuretic, whereas in the remaining 14% the diastolic pressure did not drop to the target level until a vasodilator (hydralazine) had been added to this twofold combination. Whereas, as already mentioned, a strong correlation was observed between the response to the beta-blocker and the age of the patient, there was no apparent relationship between the efficacy of the beta-blocker and the height of the blood pressure prior to treatment (cf. Figure 3). On the whole, however, the higher the pre-treatment pressure, the more often it proved necessary to administer the twofold or threefold combination; almost one out of every three patients did in fact require combined therapy.

This study has shown that, by systematically employing this triple-component treatment schedule, it is possible to "normalise" the blood pressure in more than 90% of a group of patients suffering from essential hypertension. In this type of regimen, a beta-blocker can be used as the drug of first choice. In patients over 60 years of age, however, who are usually less responsive to monotherapy with a beta-blocker, a diuretic should be given in addition from the very beginning. Treatment of this kind, based primarily on the use of a beta-blocker, is efficacious and gives rise to few unwanted effects. It also holds out the prospect of reducing cardiac morbidity and mortality – a goal which traditional antihypertensives have so far failed to achieve.

References

1 VETERANS ADMINISTRATION COOPERATIVE STUDY GROUP ON ANTIHYPERTENSIVE AGENTS: Effects of treatment on morbidity in hypertension. Results in patients with diastolic blood pressures averaging 115 through 129 mm Hg. J. Amer. med. Ass. *202*, 1028 (1967)
2 VETERANS ADMINISTRATION COOPERATIVE STUDY GROUP ON ANTIHYPERTENSIVE AGENTS: Effects of treatment on morbidity in hypertension. II. Results in patients with diastolic blood pressure averaging 90 through 114 mm Hg. J. Amer. med. Ass. *213*, 1143 (1970)

Beta-blockade in hyperkinetic heart syndrome

by M. Guazzi*

Beta-blockers exert a particularly pronounced effect in patients with hyperkinetic heart syndrome, i.e. in patients exhibiting cardiac hypersensitivity to adrenergic stimulation. Although the reasons for such hypersensitivity are still unknown, its adverse repercussions have been amply demonstrated; they may in fact sometimes be so severe as to impose an undue strain on the cardiovascular system even in the resting state – a strain reflected in elevated baseline values for the systolic and possibly also the diastolic pressure, as well as for the ventricular ejection rate and cardiac output. This haemodynamic overload is further aggravated by the emotional tension and minor anxieties that are part and parcel of normal daily life. The reduced working capacity or the left-ventricular failure which have been reported as findings in such cases might well be due to this continuous overload. In addition, evidence has been accumulating to suggest that the height of the systolic arterial pressure is directly related to the risk of coronary heart disease. Consequently, a systolic pressure in excess of 150 mm. Hg at any age may have serious clinical implications – implications involving the possible development of chronic hypertension and/or of coronary arteriosclerosis. If these potential hazards are considered serious enough to warrant appropriate therapy for hyperkinetic heart syndrome, then beta-blockers must be regarded here as an ideal form of treatment.

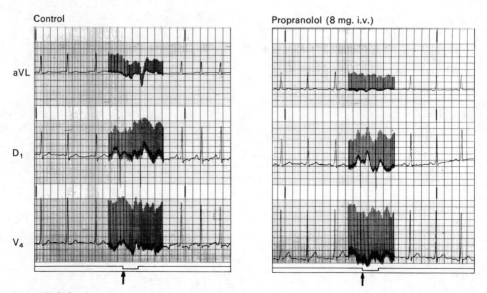

Fig. 1. E.C.G. changes occurring in response to passive head-up tilting (70°), before and after an intravenous dose of propranolol.

---

* Istituto di Ricerche Cardiovascolari dell'Università di Milano, Milan, Italy.

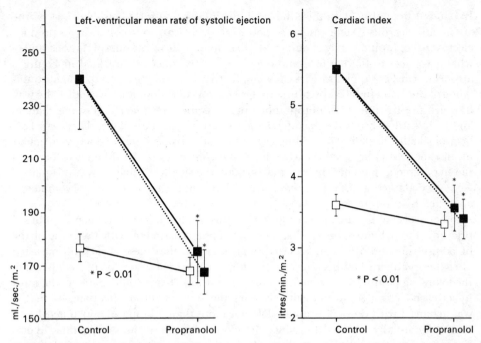

Fig. 2. Average values for left-ventricular mean rate of systolic ejection and cardiac index (per square metre of body surface), before and after treatment with propranolol in healthy subjects (□) and in patients with hyperkinetic heart syndrome (■). In each instance, the dotted line indicates the effect of propranolol when infused intravenously, and the continuous lines the effect of long-term oral treatment with propranolol.

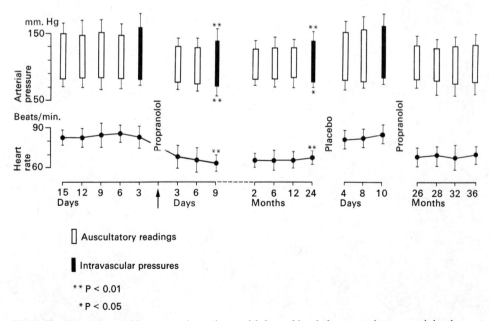

Fig. 3. Blood pressure and heart rate in patients with hyperkinetic heart syndrome receiving long-term treatment with propranolol.

In patients presenting with this syndrome, the influences exerted on the heart by the autonomic nervous system can be so powerful, or at least so peculiar, as to produce electrocardiographic changes on a scale that may lead to misinterpretations and – what is worse – to the use of unsuitable forms of treatment. Illustrated in Figure 1 are electrocardiographic tracings showing, firstly, the response to sympathetic stimulation of the heart induced by tilting under control conditions and, secondly, the way in which pre-treatment with propranolol prevents the occurrence of T-wave changes and tachycardia. From the haemodynamic standpoint, Figure 2 clearly reveals how beta-blockade normalises the resting values for both cardiac index and left-ventricular mean rate of systolic ejection: whereas in the control state the differences between the hyperkinetic patients (black squares) and the healthy subjects (white squares) were very marked and, in fact, statistically significant, they were almost completely abolished by propranolol.

Another aspect of the use of beta-blockers in this condition is their influence on blood pressure. From Figure 3 it can be seen that under treatment with propranolol the blood pressure and heart rate reverted to normal and that these effects still persisted after the treatment had been in progress for two or three years. In this connection, the following points are particularly worth noting: a) that in these patients the antihypertensive effect set in immediately after the start of the medication; b) that it was achieved with doses of propranolol (not more than 120 mg. daily) much lower than those usually required to reduce the blood pressure in cases of essential hypertension; and c) that substitution of placebo for the propranolol was promptly followed by a rise in blood pressure to pre-treatment levels. These observations, coupled with the fact that the fall in blood pressure was attributable solely to the normalisation of cardiac output, indicate that in patients with hyperkinetic heart syndrome the decrease in elevated blood pressure is directly dependent upon the drug's beta-blocking action as such and that – in contrast to other forms of hypertension treated with a beta-blocker – no other mechanisms are involved.

Summary of discussion

Prof. MUIESAN referred – in connection with the plasma renin determinations which Dr. BÜHLER had carried out in hypertensives – to the findings he himself obtained when he measured the plasma catecholamine levels in patients suffering from hypertension. An interesting feature of these findings had been the discovery that the catecholamine levels measured after physical exercise depended, not on the patients' ages, but on their state of physical fitness. Accordingly, he suggested that the general practitioner would be well advised not to base a regimen of antihypertensive therapy on individual catecholamine determinations.

Prof. GROSS agreed with this conclusion; he added that such determinations were in any case costly and time-consuming to perform and were not always accurate.

Dr. BÜHLER hastened to point out that, when presenting his paper, he had certainly not wanted to create the impression that treatment for hypertension should be based on the results of plasma catecholamine measurements. He, too, knew of data which appeared to indicate that the patient's age had no direct influence on the release of catecholamines. While agreeing that differences in the individual state of fitness of the patients did indeed pose a problem in this context, he mentioned that in his experience the release of catecholamines occurring in response to a standard "dose" of physical exercise was greater in elderly than in younger patients – a fact which, he felt, must presumably also have some influence on the blood pressure.

Prof. BIRKENHÄGER, in reply to a question put to him by Prof. GROSS, observed that, so far as he could gather from his own investigations, a modicum of scepticism concerning the value of catecholamine determinations was justified only insofar as such determinations did not seem to afford a reliable guide to adrenergic activity while a patient was actually undergoing treatment with a beta-blocker – at least not during the first two weeks of such treatment. From the fact that, in the patients he studied, the changes in the diurnal blood pressure curves did not run parallel to those in the plasma catecholamine concentrations measured during treatment with propranolol it could be provisionally concluded that propranolol inhibits the re-uptake of nor-adrenaline by the sympathetic nerve endings.

Prof. FEJFAR, having drawn attention to the increased risk of cardiovascular complications in the presence of ventricular hypertrophy, asked Prof. ZANCHETTI whether it was possible with beta-blockers to prevent or minimise ventricular hypertrophy.

Prof. ZANCHETTI thought it would be interesting and desirable to investigate this question in future studies. In this connection he emphasised how important it was that the various possible cardioprotective effects of the beta-blockers, as indicated by the results of recent research, should be weighed in the balance against other repercussions produced by these drugs on the myocardium, some of which had already been known for years.

Prof. MORET added, with reference to the query posed by Prof. FEJFAR, that in experiments which he had performed with propranolol in spontaneously hypertensive rats he had observed no effect on hypertrophy of either the right or left ventricle.

Dr. WILHELMSEN mentioned results of animal experiments carried out in FOLKOW's laboratory in Gothenburg which had shown that it was in fact possible to prevent left-ventricular hypertrophy in spontaneously hypertensive rats by treating them with beta-blockers from birth onwards; no such preventive effect was observed, however, when this treatment was initiated after the first six months of life.

Dr. SOMERVILLE suggested that, with regard to the cardioprotective action of beta-blockade, caution was indicated on two scores:

Firstly, he felt that, despite the very encouraging preliminary findings obtained to date, the relatively wide gap separating the undoubtedly interesting and attractive theory of cardioprotection with beta-blockers from the realities of clinical practice had yet to be bridged. Genuine proofs confirming this cardioprotective activity could be counted on the fingers of one hand, and for this purpose one wouldn't even need to use all five fingers! The results of several of the trials undertaken with beta-blockers in the United Kingdom had not been conducive to enthusiasm, and clinical proof of a cardioprotective effect *in the acute phase of myocardial infarction* was still lacking. On the other hand, the evidence in favour of using beta-blockers *for the prevention of infarction* was possibly more convincing.

Secondly, it seemed to him that Dr. BÜHLER's paper might create the impression that the problem of treatment for hypertension had already been solved and that, depending on the case, it was now possible – either with monotherapy or with a twofold or threefold combination – to lower every patient's elevated blood pressure. Unfortunately, this was not so. In numerous cases, the blood pressure remained excessively high despite all attempts to lower it; in this context, he was thinking of hypertensives exhibiting an elevated peripheral resistance which proved refractory to treatment; in such patients one could admittedly reduce cardiac output, but only to a certain extent. What, he wondered, would Dr. BÜHLER do about the problem of these refractory cases?

Prof. GROSS reminded the audience at this point that at the beginning of the symposium he had tried to make it clear that cardioprotection by beta-blockade was a "simplifying slogan" and that the individual effects covered by the term required to be defined. In this context, one should above all beware of indulging in oversimplifications.

Dr. BÜHLER acknowledged that among his patients, too, he also encountered cases which were refractory to treatment. The fact remained, however, that of 155 patients suffering from uncomplicated essential hypertension 142 had responded satisfactorily to the regimens he had described. It was thus possible to achieve successful results in 90–95% of such hypertensives by giving them monotherapy with either a beta-blocker or a diuretic or by treating them with these two drugs in combination, plus the vasodilator hydralazine as and where necessary.

Dr. TAYLOR remarked that in refractory cases of hypertension it was usually only the resting blood pressure that could not be lowered, or at least not lowered sufficiently. But, even in these patients, use of a beta-blocker did at all events serve to prevent marked additional rises in blood pressure occurring in response to physical exertion.

Dr. SEVER suggested that the type of patient treated by Dr. BÜHLER might have had some bearing on the good results he obtained. In his own experience, coloured patients from the West Indies or Africa, of whom there were many in London, did not respond very well to treatment with beta-blockers.

Dr. BÜHLER mentioned in reply to Dr. SEVER that he had previously also carried out studies in New York which had likewise confirmed that in Negroes it was more difficult to obtain an adequate fall in blood pressure by administering monotherapy with a beta-blocker.

Dr. Török commented that in patients suffering from moderate hypertension she had been unable to detect any significant correlation between the extent of the fall in blood pressure produced by a beta-blocker and either the pre-treatment cardiac index or the subsequent decrease in the cardiac index. In patients with hyperkinetic heart syndrome, by contrast, the cardiac index had decreased *pari passu* with the blood pressure right from the start of treatment with propranolol, and this decrease had also persisted throughout the duration of long-term therapy lasting two years.

Prof. Guazzi thought it a pity that in hypertensives it was not possible in general to make any predictions about the success or failure of beta-blocker therapy on the basis of the patient's haemodynamic profile.

Prof. Zanchetti drew attention in this connection to studies undertaken by Tarazi and his team, who had also found it impossible to predict from haemodynamic data how their patients would respond to antihypertensive treatment with propranolol.

Prof. Guazzi added that the notion that beta-blockers are particularly indicated in hypertensives with an elevated cardiac output was probably an overhasty conclusion; he thought it even conceivable that in cases of essential hypertension it might, on the contrary, be precisely patients with an elevated peripheral resistance who responded better.

Prof. Birkenhäger could not agree with this suggestion. Judging from the results he himself had obtained, there appeared to be no correlation at all between a patient's response to beta-blockade and either his cardiac output or his total peripheral resistance.

Prof. Wollheim observed that among his own patients there was a relatively small group of youngish hypertensives exhibiting an elevated cardiac output. He did not think, however, that these younger patients were necessarily all suffering from essential hypertension or would necessarily remain hypertensive.

With reference to Dr. Bühler's paper, he concurred that beta-blockers were especially indicated in younger patients. Cardiac performance deteriorated with age – not only in hypertensives but also in the population as a whole. For this reason he adopted a very cautious attitude towards the use of beta-blockers in elderly hypertensives, in whom the possible adverse repercussions of beta-blockade on the heart might well outweigh such benefit as a cardioprotective effect could confer.

Dr. Bühler, while agreeing in principle with these comments, pointed out that the same argument also applied to most conventional antihypertensives – such as reserpine, clonidine, and methyldopa – which were also capable of diminishing cardiac output.

Beta-blockers in the treatment of angina pectoris: the prevention of myocardial infarction

by J.-L. RIVIER*

As far as their ability to combat pain is concerned, there is no longer any doubt about the effectiveness of beta-blockers in patients with angina pectoris[22]. They afford marked relief from pain in 70–80% of such cases, provided the dosage in which they are administered is gradually raised to the requisite level and does not provoke signs of intolerance. This success rate can be improved upon still further by selecting the best-tolerated beta-blocker and by giving it in combination with nitrates[25].

The question that now arises is whether the beta-blockers can influence the natural course of angina pectoris, i.e. whether they can prolong the patient's life expectancy and reduce the risk of infarction.

The mean annual mortality rate for patients with angina pectoris is 4%[14], but this figure really only applies to large series of cases. The course of the disease does in fact vary appreciably, depending on the number of main coronary arteries that have suffered serious damage (reduction of 50% or more in the vascular lumen) and on the status of the left ventricle. According to Ross[23], who recently published a critical review of the entire literature on the subject, the mean annual mortality rate is 10% for cases in which two or three main coronary arteries are affected, and 2% or less if only one main artery is involved. The functional status of the ventricles plays a major role in the prognosis, and the greater the degree of its impairment, the poorer the prognosis[4]: in the presence of heart failure, the annual mortality rate rises to 12.4%, while in patients in whom ventriculography reveals signs of asynergy or whose haemodynamic patterns are abnormal it amounts to approximately 8%. Arterial hypertension is likewise a factor of considerable prognostic significance, since the annual mortality rate for angina pectoris patients with hypertension also reaches 8%[4]. Angina pectoris *per se,* on the other hand, has no influence on the prognosis[4,11]. In patients with ischaemic heart disease confirmed by coronary arteriography BURGGRAF and PARKER[4] have calculated the following annual mortality rates: 4.8% for cases without angina; 5.8% for cases with exertional angina; 5.9% for cases with angina at rest; and 6% for cases with a past history of myocardial infarction. The differences are thus not significant. Leaving cases of "unstable" angina pectoris out of account, the severity of the pain likewise has little (if any[4]) effect on the prognosis, the annual mortality rate being 6% for angina pectoris of Grades 1 and 2 and 8% for Grades 3 and 4[4,11,20]. The duration of the angina, however, does have a bearing on the prognosis: the annual mortality rate over a five-year period in cases of stable angina is reported to be 4–5.8% where the disease has been present for 1–6 years and 10% where it has persisted for more than six years[20].

From these figures the following conclusions can be drawn: the annual mortality rate for angina pectoris is roughly the same as for asymptomatic coronary artery disease with or without a past history of myocardial infarction – i.e. it is about 4–6%.

---

* Division de cardiologie du Département de médecine, Centre hospitalier universitaire vaudois, Lausanne, Switzerland.

This percentage should be borne in mind, because in large series of patients suffering from various grades of angina pectoris treatment with beta-blockers would have to reduce this figure by a significant amount before it could be held to be of proven therapeutic value. In studies involving relatively small numbers of patients the selection of cases has a very pronounced influence on the results achieved, and, in my opinion, this selection – at least as far as patients with no past history of myocardial infarction are concerned – should be based in part on findings obtained by coronary arteriography.

Since the prognosis in coronary artery disease is virtually the same whether or not angina pectoris is also present, it is not necessary to include angina patients in controlled, randomised studies with beta-blockers in order to assess the efficacy of these drugs. This is to be welcomed from the ethical point of view, because in many instances there is, at the moment at least, no substitute for the beta-blockers as symptomatic therapy.

The possibility that the beta-blockers may exert a beneficial influence on the course of angina pectoris is all the more attractive since no other effective means of combating coronary arteriosclerosis itself and of preventing sudden death or re-infarction is available. Neither anticoagulants[7] nor antilipidaemic agents[5] nor inhibitors of platelet aggregation can be depended upon to serve this purpose, although there is some hope that inhibitors of platelet aggregation might prove useful[12]. Surgical treatment specifically designed to achieve direct revascularisation of the myocardium by means of an aorto-coronary bypass occupies a special position; increasing evidence has been accumulating to show that this operation can be effective under certain circumstances – for example, in the presence of stenosis of the left main coronary artery[27] and in some severe forms of unstable angina pectoris[17]. Whether it is of any value for the majority of cases, however, remains a moot point[4], even if it does produce a symptomatic improvement in 80–90% of those patients in whom it is performed. Moreover, the operation has a merely palliative effect and can only be considered in cases where the coronary artery occlusions are favourably located; hence, an aorto-coronary bypass is to be regarded simply as a supplement to long-term drug therapy, and never as a replacement for it, even if the patient does become symptom-free as a result. It is important to emphasise this point in any discussion of the use of beta-blockers for the long-term treatment of coronary artery disease.

The results obtained in five studies[1, 6, 16, 26, 28] in which the efficacy of beta-blockers in angina pectoris was assessed over varying periods of time are shown in Table 1;

Table 1. Beta-blockers in the treatment of angina pectoris: effects on annual mortality.

Authors	Year	Number of cases	Coronary arteriography	Duration of study	Annual mortality
Amsterdam et al.[1]	1968	43	+	1.5 years	3.8%
Lambert[16]	1974	139	–	3 years	1.6%
Russek[26]	1975	103*	±	5 years	1.2%
Faerchtein et al.[6]	1974	24	–	2 years	2%
Warren et al.[28]	1976	63	+	3.8 years	3.8%
(Russek[26]	1975	31**	±	4 years	25%)

* Patients with a good prognosis
** Patients with a poor prognosis

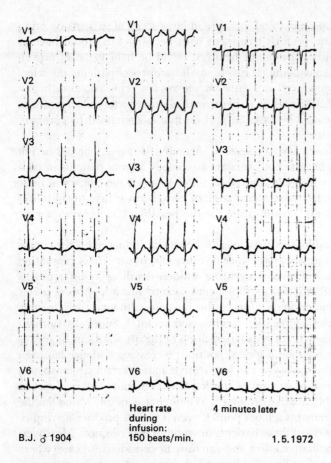

Fig. 1. Effect of an isoprenaline infusion on the electrocardiogram of a patient with coronary artery disease.

the mean annual mortality rate in the various series of patients (excluding those with a poor prognosis in the study by RUSSEK[26]) was 2.5%, the minimum being 1.2% and the maximum 3.8%. Since the mortality rate for untreated cases of angina pectoris is 4%, some of the figures reported in Table 1 seem promising. According to RUSSEK, the five-year prognosis is good even for patients with severe damage (reduction of 70% or more in the vascular lumen) affecting three main coronary arteries; in these cases the annual mortality rate was 2.4%, provided left-ventricular status was good and optimum treatment with beta-blockers and nitrates was given. All these findings are certainly encouraging.

A further problem to be considered is the mechanism by which the beta-blockers could possibly reduce both the mortality rate and the risk of infarction in patients with angina pectoris. Animal-experimental studies in which beta-blockers were used as a means of protecting the myocardium during the acute phase of an infarction may perhaps shed some light on this point. Leaving aside the question as to the relative merits of the various experimental procedures employed in these studies, it is agreed by all the investigators that beta-blockers – or, to be more exact, propranolol, since this was the drug almost exclusively utilised – are in fact capable of reducing the size of an experimental infarction and limiting the consequences of acute ischaemia following temporary occlusion of a coronary artery[2,18]. Is this finding to be ascribed to the fact that beta-blockade modifies the distribution of the coronary blood supply

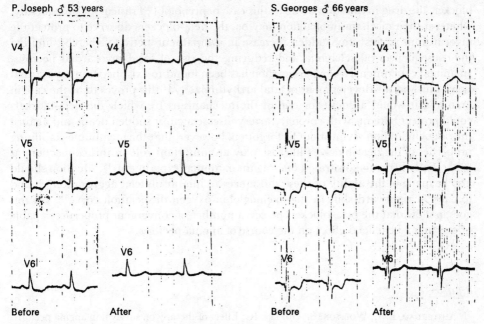

Fig. 2. Effect of isoprenaline on the electrocardiogram of two patients with coronary artery disease before and after treatment with oxprenolol.

in such a way that irrigation of the subendocardial layers and of the ischaemic zones is actually improved? Some authors claim to have demonstrated such redistribution in the dog[2], while others failed to observe it in either guinea-pigs[8] or dogs[21]. No firm answer to this question can therefore be given until more information is available. Nevertheless, it is already clear that propranolol affords the myocardial cells protection in the presence of ischaemia, inasmuch as following treatment with this drug the cells' content of certain substances rich in energy, such as creatine phosphate, is less reduced than in controls and their lactate concentration less elevated[3]. The fact that this beneficial effect of propranolol is linked to the drug's beta-blocking action, and not to its membrane-stabilising properties (D-propranolol being ineffective in this respect[21]), suggests that beta-blockers act directly on the myocardial cell by attenuating metabolic disturbances which are due to lack of oxygen and in which the beta-receptors are involved[21]. If this also applies to man, then the effect of propranolol in this instance would have to be attributed, not so much to the well-known haemodynamic changes which it produces and which lead to a decrease in myocardial oxygen consumption, as to blockade of the receptors in the cell itself. All sympathetic stimulation, including in particular that induced by the catecholamines, would thus be blocked. It has indeed been convincingly shown in animals that the catecholamines have a deleterious effect on the myocardium, chiefly because they eventually give rise to small foci of necrosis even in the absence of a reduction in coronary blood flow[10,13]. In man, too, the harmful effects of the catecholamines on the myocardium have repeatedly been demonstrated: in patients with coronary artery disease the infusion of isoprenaline, for example, induces not only an increase in heart rate, but also, subsequently, characteristic electrocardiographic changes due to ischaemia (Figure 1). As shown in Figure 2, these changes do not occur following pre-treatment with a beta-blocker, the drug in question being in this case oxprenolol (®Trasicor). Consequently, whatever their mechanism of action may be, the beta-blockers do serve to protect the myocardium against any sudden increase in sympathetic activity, thereby limiting the consequences of ischaemia and reducing the risk of focal myocardial necrosis. In animal experiments a close association has been found to exist between myocardial ischaemia and the development of fatal arrhythmias[29]. If this proves to be the case in man too, it would account for one of the mechanisms by which the beta-blockers could protect patients with coronary artery disease against sudden death and against the risk of infarction. It must not be forgotten, however, that the membrane-stabilising properties of the beta-blockers likewise play an important role in this connection, as has recently been demonstrated once again in animal experiments[15]. Mention should also be made of the ability of these substances to inhibit platelet aggregation[9, 24] and to cause a rightward shift in the haemoglobin-oxygen dissociation curve[19]. All this goes to show that the beta-blockers possess a number of convergent properties capable of exerting a beneficial effect on the course of angina pectoris.

References

1 AMSTERDAM, E.A., WOLFSON, S., GORLIN, R.: Effect of therapy on survival in angina pectoris. Ann. intern. Med. *68*, 1151 (1968); abstr.
2 BECKER, L.C., FORTUIN, N.J., PITT, B.: Effect of ischemia and antianginal drugs on the distribution of radioactive microspheres in the canine left ventricle. Circulat. Res. *28*, 263 (1971)
3 BRACHFELD, N.: Metabolic evaluation of agents designed to protect the ischemic myocardium and to reduce infarct size. Amer. J. Cardiol. *37*, 528 (1976)
4 BURGGRAF, G.W., PARKER, J.O.: Prognosis in coronary artery disease: angiographic, hemodynamic and clinical factors. Circulation *51*, 146 (1975)
5 THE CORONARY DRUG PROJECT RESEARCH GROUP: Clofibrate and niacin in coronary heart disease. J.Amer. med. Ass. *231*, 360 (1975)
6 FAERCHTEIN, I., ROQUE, A.F., KASTANSKY, L.: Long-term treatment of angina pectoris with MK-950. In Magnani, B. (Editor): Beta-adrenergic blocking agents in the management of hypertension and angina pectoris (Raven Press, New York 1974)
7 FEINSTEIN, A.R.: More blood for the anticoagulant battle. New Engl. J. Med. *292*, 1400 (1975)
8 FRICK, M.H., VIRTANEN, K.S.: Beta-blockade and coronary circulation. Amer. Heart J. *91*, 536 (1976)
9 FRISHMAN, W.H., WEKSLER, B., CHRISTODOULOU, J.P., SMITHEN, C., KILLIP, T.: Reversal of abnormal platelet aggregability and change in exercise tolerance in patients with angina pectoris following oral propranolol. Circulation *50*, 887 (1974)
10 HAYASE, S., ITO, H., KONDO, Y., et al.: Inhibition action of propranolol and its stereoisomers on epinephrine-induced changes in electrocardiogram. Jap. Circulat. J. *36*, 1065 (1972)
11 HUMPHRIES, J.O., KULLER, L., ROSS, R.S., FRIESINGER, G.C.: Natural history of ischemic heart disease in relation to arteriographic findings. Circulation *49*, 489 (1974)
12 JICH, H.: Aspirin and myocardial infarction. Amer. Heart J. *91*, 126 (1976)
13 KAHN, D.S., RONA, G., CHAPPEL, C.: Isoproterenol-induced cardiac necrosis. Ann. N.Y. Acad. Sci. *156*, 285 (1969)
14 KANNEL, W.B., FEINLAB, M.: Natural history of angina pectoris in the Framingham study. Prognosis and survival. Amer. J. Cardiol. *29*, 154 (1972)
15 KUPERSMITH, J., SHIANG, H., LITWAK, R.S., HERMAN, M.V.: Electrophysiological and antiarrhythmic effects of propranolol in canine acute myocardial ischemia. Circulat. Res. *38*, 302 (1976)
16 LAMBERT, D.M.D.: Long-term survival on beta-receptor-blocking drugs in general practice – a three-year prospective study. In Burley, D. M., et al.(Editors): Hypertension – its nature and treatment, Int. Symp., Malta 1974, p. 283 (CIBA Horsham, England, 1975)
17 LAWSON, R.M., CHAPMAN, R., WOOD, J., STARR, A.: Acute coronary insufficiency – an urgent surgical condition. Brit. Heart J. *37*, 1053 (1975)

18 MAROKO, P.R., KJEKSHUS, J.K., SOBEL, B.E., WATANABE, T., COVELL, J.W., ROSS, J., Jr., BRAUNWALD, E.: Factors influencing infarct size following experimental coronary artery occlusions. Circulation *43*, 67 (1971)
19 OSKI, F.A., MILLER, L.D., DELIVORIA-PAPADOPOULOS, M., MANCHESTER, J.H., SHELBURNE, J.C.: Oxygen affinity in red cells: changes induced in vivo by propranolol. Science *175*, 1372 (1972)
20 PROUDFIT, W.L.: Prognostic factors in atherosclerotic coronary disease. In Vidt, D.G. (Editor): Cleveland clinic cardiovascular consultations, p. 19; Cardiovascular clinics (Davis, Philadelphia 1975)
21 REIMER, K.A., RASMUSSEN, M.M., JENNINGS, R.B.: On the nature of protection by propranolol against myocardial necrosis after temporary coronary occlusion in dogs. Amer. J. Cardiol. *37*, 520 (1976)
22 RIVIER, J.-L.: Beta-blockers in the treatment of angina pectoris. In Schweizer, W. (Editor): Beta-blockers – present status and future prospects, Int. Symp., Juan-les-Pins 1974, p. 193 (Huber, Berne/Stuttgart/Vienna 1974)
23 ROSS, R.S.: Ischemic heart disease: an overview. Amer. J. Cardiol. *36*, 496 (1975)
24 RUBEGNI, M., PROVVEDI, D., BELLINI, P.G., BANDINELLI, C., DE MAURO, G.: Propranolol and platelet aggregation. Circulation *52*, 964 (1975); corresp.
25 RUSSEK, H.I.: Propranolol and isosorbide dinitrate synergism in angina pectoris. Amer. J. Cardiol. *21*, 44 (1968)
26 RUSSEK, H.I.: Prognosis in angina pectoris with optimal medical therapy. In Russek, H.I. (Editor): New horizons in cardiovascular practice, p. 151 (University Park Press, Baltimore/London/Tokyo 1975)
27 TAKARO, T., HULTGREN, H.N., DETRE, K.M.: VA cooperative study of coronary arterial surgery: II. Left main disease. Circulation *51/52*, Suppl. II: 143 (1975)
28 WARREN, S.G., BREWER, D.L., ORGAIN, E.S.: Long-term propranolol therapy for angina pectoris. Amer. J. Cardiol. *37*, 420 (1976)
29 WIT, A.L., BIGGER, J.T.: Possible electrophysiological mechanism for lethal arrhythmias accompanying myocardial ischemia and infarction. Circulation *51/52*, Suppl. III: 96 (1975)

Influence of beta-blockers on the incidence of re-infarction and sudden death after myocardial infarction

by C. Wilhelmsson, A. Vedin, and L. Wilhelmsen *

Introduction

In Western countries cardiovascular diseases are the commonest cause of death[9,13] and coronary heart disease is responsible for approximately two-thirds of the cardiovascular mortality. In acute episodes of myocardial infarction most patients die suddenly outside hospital, and the major predictor of such sudden deaths is a previous myocardial infarction in the case history[14].
Since 1968 results have been presented suggesting that beta-receptor blockade exerts a favourable effect following myocardial infarction. Although studies in man have not yet convincingly demonstrated that treatment with a beta-blocker can reduce the mortality from acute myocardial infarction, since 1974 several long-term studies of post-infarction patients have shown that a reduction in the number of sudden deaths occurring in these patients can be achieved by subjecting them to treatment with beta-blockers[1,4,18].

The Gothenburg study

Included in this Gothenburg study[16,18] were all men and women aged 57–67 years who had suffered an acute myocardial infarction and had afterwards been discharged from hospital during the years 1970 and 1971. As can be seen from Table 1, 114 of these patients were treated with 400 mg. alprenolol (®Gubernal) daily, and 116 with placebo. The patients were followed up at a special clinic for post-infarction cases, where standardised indications for symptomatic therapy were adopted and standardised criteria applied for the detection of cardiovascular risk factors and complications[3]. Upon admission to the trial the patients were divided into four prognostically homogeneous subgroups designated I to IV[14] (Figure 1). After two years the number of sudden deaths showed a significant reduction in Subgroups II and IV. No deaths occurred in Subgroups I and III. The difference between the results obtained in the

Table 1. Number of patients allocated to treatment with alprenolol and placebo, respectively, in the four risk groups (Gothenburg study).

Risk group	Alprenolol (400 mg.)	Placebo
I	28	29
II	54	57
III	4	3
IV	28	27
Total	114	116

* Medicinska kliniken I, Sahlgrenska sjukhuset, Gothenburg, Sweden.

patients treated with alprenolol and those receiving the placebo was statistically significant (Table 2 and Figure 2), the total mortality being only half as high in the patients under treatment with the beta-blocker. This was true regardless of whether the groups were analysed *per se* or were pooled in various ways. There was no difference between the two treatments with respect to non-fatal re-infarction, which occurred in 16 of the alprenolol-treated patients and in 18 of those treated with the placebo (Table 2 and Figure 3). The drop-out rate was low: only 16 patients – eight on alprenolol and eight on the placebo – dropped out during the study. Patient compliance was also monitored; tablet counting and urine analyses showed that 80% of the patients had been taking at least 90% of the tablets prescribed.

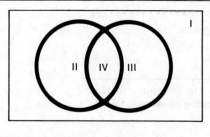

Risk group I   *No extensive cardiac damage*

Risk group II   *Mechanical damage to myocardium*
1) Relative heart volume ♂ > 450 ml./m.² body surface area
                                  ♀ > 400 ml./m.² body surface area
2) Glutamic pyruvic transaminase >40 U. during the first 3 days
3) Body temperature >38 °C. during the first 3 days
4) Transient atrial flutter and fibrillation

Risk group III   *Electrical cardiac damage*
1) Ventricular premature beat: frequency >5/min.
2) Ventricular tachycardia or ventricular fibrillation
3) Atrioventricular block:
    a. PQ >0.24 sec.
    b. Type II
    c. Type III

Risk group IV   *Combined electromechanical damage*
                    (as in Risk groups II/III)

Fig. 1. Criteria adopted for assigning patients to the four homogeneous risk groups (Gothenburg study).

Table 2. Number of non-fatal re-infarctions and sudden deaths, as well as total number of deaths, per risk group recorded after treatment with alprenolol or placebo (Gothenburg study).

	Alprenolol Risk group				Placebo Risk group			
	I	II	III	IV	I	II	III	IV
Non-fatal re-infarctions	4	9	0	3	4	7	0	7
Sudden deaths	0	1	0	2	0	6	0	5
Total number of deaths	0	4	0	3	0	7	0	7
Number of patients included in the evaluation*	25	52	3	26	27	55	3	23

* Patients who during the course of the trial developed contra-indications to continued participation have been excluded.

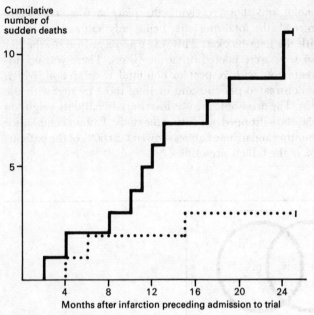

Fig. 2. Cumulative number of sudden deaths in the group treated with alprenolol and the placebo group during the two-year follow-up period (Gothenburg study).

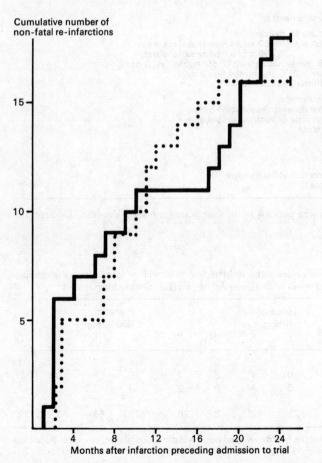

Fig. 3. Cumulative number of non-fatal re-infarctions in the group treated with alprenolol and the placebo group during the two-year follow-up period (Gothenburg study).

The international multicentre study

The international multicentre study involved 3,038 patients from 67 centres, each of which contributed between 4 and 162 patients (Table 3). The patients participating in this large study, who had all been discharged from hospital 7–28 days after having suffered an acute myocardial infarction, were randomly assigned to treatment with either practolol (400 mg. daily) or placebo. The practolol-treated group showed a significant reduction in the number of sudden deaths (Table 4). Although it is impossible to form an opinion about the representativeness of the patients, this does not detract from the value of the main results of the study, included among which was also a significant decrease in the incidence of "all cardiac events". On the other hand, the retrospective demonstration that patients with anterior wall infarcts and low diastolic blood pressures do particularly well on practolol seems questionable, and the reliability of findings based on such a discriminant analysis requires prospective validation; at present there are thus no grounds for concluding that beta-blockade is especially favourable in these specific cases.

Other studies

Evidence in support of the observations reported above has also emerged from an open study, performed in Falun (Sweden), in which post-infarction patients were

Table 3. Numbers of patients at risk (international multicentre study[4]).

	Patients admitted to the trial		Patients withdrawn, died, or dropped out at each stage	
	Practolol	Placebo	Practolol	Placebo
At start of trial	1,524	1,514		
After 1 month	1,427	1,418	97	96
After 3 months	1,356	1,325	168	189
After 6 months	1,286	1,236	238	278
After 12 months	1,006	971	518	543
After 18 months	562	547	962	967
After 24 months	330	336	1,194	1,178

Mean duration of participation in the trial: patients on practolol 14.3 months; patients on placebo 14.0 months.

Table 4. Total number of deaths and re-infarctions (international multicentre study[4]).

	Practolol	Placebo	Statistical significance (P)
Cardiac deaths	47	73	< 0.02
Non-fatal re-infarctions	69	89	< 0.10
All cardiac events	116	162	< 0.01
Non-cardiac deaths	5	3	
Cardiac deaths after withdrawal from the trial	36	37	
Non-cardiac deaths after withdrawal from the trial	6	4	

treated for two years with 400 mg. alprenolol daily[1]. In this study, 69 alprenolol-treated patients were compared with 93 control patients. Other drugs, where indicated, were given in accordance with conventional criteria. In the alprenolol group there was only one occurrence of sudden death, whereas in the placebo group there were nine sudden deaths (Figure 4). In the alprenolol group the number of patients with re-infarction was also significantly lower (Figure 5). The total mortality in the

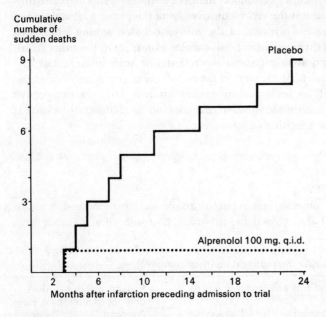

Fig. 4. Cumulative number of sudden deaths in the group treated with alprenolol and the placebo group during the two-year follow-up period (open study performed in Falun[1]).

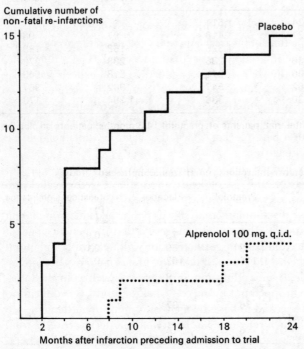

Fig. 5. Cumulative number of non-fatal re-infarctions in the group treated with alprenolol and the placebo group during the two-year follow-up period (open study performed in Falun[1]).

alprenolol group worked out at 7%, as compared with 13% in the control group. In a retrospective analysis the two groups were found to be fully comparable at the start of the study with respect to a number of prognostically relevant variables.

The situation as regards the prophylactic value of treatment with beta-blockers in patients who have not yet had an infarction is still unclear. Retrospective studies undertaken in Great Britain suggest that beta-blockers improve the life expectancy of patients with angina pectoris and hypertension [6, 7]. However, in the case of patients with hypertension and/or angina pectoris, as well as in other categories of patient in which the risk of acute coronary disease is high, these British findings have yet to be corroborated by controlled trials. Incidentally, several clinical observations have indicated that the mortality, or the risk of infarction, increases following the sudden withdrawal of beta-blockers in patients receiving these drugs as treatment for angina pectoris [2, 12].

The risk of fatal and non-fatal re-infarction is drastically increased among patients who have already suffered one myocardial infarction. During the 12 months following their first infarction the risk of death and re-infarction in these patients is 30 times higher than in healthy subjects. That the possibilities of preventing sudden death by recourse to drug treatment should first have been investigated in post-infarction patients is therefore readily understandable.

Comments

During recent years, results of animal studies have been published which indicate that beta-blockade has a favourable effect on myocardial metabolism, resulting in a decrease in the extent of experimentally induced infarctions [8, 10, 11]. The essential difference between long-term post-infarction studies of the type described here and previous investigations concerned with the use of beta-blockers as a means of preventing cardiac arrhythmias in cases of acute myocardial infarction lies in the fact that the patients on long-term treatment already had a therapeutic blood concentration of the beta-blocker at the time of the fresh episode, whereas in the acute studies the drug was given only after the infarction had actually occurred. The findings obtained in these long-term trials tally well with the results of animal experiments [5]. In other words, evidence from independent sources has now demonstrated that long-term treatment with alprenolol and practolol after a myocardial infarction can prevent sudden deaths. It is still a moot point, however, as to whether all the patients in question should be treated with a beta-blocker or whether such treatment should be restricted to certain high-risk categories.

In the above-mentioned study carried out in Gothenburg a prognostic system was applied prospectively with regard to the risk of death; the patients were divided into four different subgroups, in two of which no deaths occurred. In the international multicentre study a retrospective discriminant analysis was performed after completion of the study; the findings disclosed by this analysis suggested that the beneficial effect of practolol was most pronounced in patients with anterior wall infarction and initial diastolic pressures which were below the mean. But, owing to shortcomings both in the data on the patients at the time of their admission to the trial, as well as in the information available on those who dropped out in the course of the study, it is difficult to interpret the results of the analysis and to base any generalisations upon them [15].

In the three clinical studies to which reference has been made, a decrease in the total number of "all cardiac events" was demonstrated. It is therefore possible that beta-blockers may serve not only to reduce the risk of sudden death but also to prevent myocardial re-infarction. The methods available today for forecasting cardiac events are mainly effective in predicting death. The likelihood of a non-fatal recurrence is generally not related to the same risk factors, and non-fatal re-infarction has thus been found difficult to predict[17]. The fact remains, however, that a recurrence which does not prove immediately fatal nevertheless constitutes a major factor in worsening the overall long-term prognosis[16].

Under these circumstances the following conclusions can be drawn, depending on whether or not one regards it as advisable to employ beta-blockers as prophylaxis against sudden death in patients who have had a myocardial infarction:

If it is considered necessary to institute long-term treatment with a beta-blocker following an infarction, then it is impossible to separate out any one particular category of patients in which such treatment would not be indicated; the reason for this conclusion is our inability to predict non-fatal re-infarction with any degree of reliability. On the other hand, if it is felt that the benefit to be derived from beta-blockade following an infarction has not yet been sufficiently well documented, then a controlled trial may be justified. If the patients participating in such a trial are grouped prognostically with regard to the risk of death, the informative value of the findings will be correspondingly greater and the high-risk group can be expected to yield results within a correspondingly shorter space of time. The trial would have to be closely monitored, however, and interrupted at once if the rate of all cardiac events among the placebo-treated groups as a whole showed a significant increase as compared with the rate in the groups receiving the beta-blocker.

References

1 AHLMARK, G., SAETRE, H., KORSGREN, M.: Reduction of sudden deaths after myocardial infarction. Lancet *1974/II,* 1563; corresp.
2 DIAZ, R.G., SOMBERG, J., FREEMAN, E., LEVITT, B.: Withdrawal of propranolol and myocardial infarction. Lancet *1973/I,* 1068; corresp.
3 ELMFELDT, D., WILHELMSEN, L., TIBBLIN, G., VEDIN, A., WILHELMSSON, C., BENGTSSON, C.: Post-infarct clinic. Follow-up of myocardial infarction patients in a specialized out-patient clinic. Acta med. scand. *197,* 497 (1975)
4 GREEN, K.G., et al.: Improvement in prognosis of myocardial infarction by long-term beta-adrenoreceptor blockade using practolol. A multicentre international study. Brit. med. J. *1975/III,* 735
5 KELLIHER, G.J., WIDMER, C., ROBERTS, J.: Seventh annual meeting of the international study group for research in cardiac metabolism. Quebec 1974; (abstr. 141)
6 LAMBERT, D.M.D.: Beta-blockers and life expectancy in ischaemic heart disease. Lancet *1972/I,* 793; corresp.
7 LAMBERT, D.M.D.: Hypertension and myocardial infarction. Brit. med. J. *1974/III,* 685; corresp.
8 LEINBACK, R.C., COLD, H.K., BUCKLEY, M.J., AUSTEN, G.W., SANDERS, C.A.: Reduction of myocardial injury during acute myocardial infarction by early application of intraaortic balloon pumping and propranolol. Circulation *48,* Suppl. IV:100 (1973); abstr.
9 LEW, E.A., ENTMACHER, P.S.: Mortality in cardiovascular disease. In Waldenström, Larson, Ljungstedt (Editors): Early phases of coronary heart disease, p.67 (Skandia International Symposia. Nordiska Bokhandel, Stockholm 1973)
10 MAROKO, P.R., LIBBY, P., COVELL, J.W., SOBEL, B.E., ROSS, J., Jr., BRAUNWALD, E.: Precordial ST segment mapping: an atraumatic method for assessing the extent of myocardial ischemic injury. The effects of pharmacologic and hemodynamic interventions. Amer. J. Cardiol. *29,* 223 (1972)

11 PELIDES, L.J., REID, D.W., THOMAS, M., SHILLINGFORD, J.P.: Inhibition by beta-blockade of the ST segment elevation after acute myocardial infarction in man. Cardiovasc. Res. 6, 295 (1972)
12 SLOME, R.: Withdrawal of propranolol and myocardial infarction. Lancet *1973/I,* 156; corresp.
13 VEDIN, J.A., WILHELMSSON, C., BOLANDER, A.M., WERKÖ, L.: Mortality trends in Sweden 1951–1968 with special reference to cardiovascular causes of death. Acta med. scand. *189,* Suppl. 515:1 (1971)
14 VEDIN, J.A., WILHELMSSON, C., ELMFELDT, D., WILHELMSEN, L., WERKÖ, L.: Sudden death: identification of high risk groups. Amer. Heart J. *86,* 124 (1973)
15 VEDIN, J.A., WILHELMSSON, C.E.: Long-term postinfarction treatment with practolol. Brit. med. J. *1975/IV,* 579; corresp.
16 VEDIN, A., WILHELMSSON, C., TIBBLIN, G., WILHELMSEN, L., WERKÖ, L.: Design of the study. In Vedin, A., et al.: Chronic alprenolol treatment of patients with acute myocardial infarction after discharge from hospital. Acta med. scand. *197,* Suppl. 575:9 (1975)
17 VEDIN, A., WILHELMSEN, L., WEDEL, H., PETTERSSON, B., WILHELMSSON, C., ELMFELDT, D., TIBBLIN, G.: Prediction of cardiovascular deaths and non-fatal reinfarctions after myocardial infarction. Acta med. scand. (printing)
18 WILHELMSSON, C., VEDIN, J.A., WILHELMSEN, L., TIBBLIN, G., WERKÖ, L.: Reduction of sudden deaths after myocardial infarction by treatment with alprenolol. Preliminary results. Lancet *1974/II,* 1157

Circulatory changes after beta-blockade in patients with coronary artery disease

by A. Reale and M. Motolese *

Beta-blockers are now being extensively used in the treatment of angina pectoris, and it has been shown that these drugs exert a protective action by helping to prevent myocardial infarction, re-infarction, and sudden death[1-3, 7, 9]. However, since coronary artery disease is a frequent cause of myocardial dysfunction, it has been stressed that in the presence of such disease propranolol may have a deleterious effect upon cardiac performance[4-6, 8]. In certain patients at least, this would, of course, impose limitations on the use of beta-blockers.

With a view to shedding light on this problem, we undertook a study in which we employed two beta-blockers with different pharmacological properties: metoprolol (®Lopresor), a cardioselective beta-blocker displaying no intrinsic sympathomimetic activity, and oxprenolol (®Trasicor), a non-cardioselective beta-blocker which does possess intrinsic sympathomimetic activity.

The investigation was carried out in two groups of untreated, non-premedicated patients with coronary artery disease undergoing diagnostic cardiac catheterisation (11 patients in each group). Haemodynamic variables were recorded before and 40 minutes after intravenous administration of either 10 mg. metoprolol or 10 mg. oxprenolol. In ten patients in each group supine exercise was also performed before and after beta-blockade.

The individual results obtained at rest are shown in Tables 1 and 2. In Table 3 the findings are expressed as mean changes from pre-drug control values for left-ventricular end-diastolic pressure (LVEDP), contractility (LVdp/dt and $\frac{LVdp/dt}{IP}$). and compliance ($\Delta P/\Delta V$), cardiac index (CI), stroke volume index (SVI), and heart rate (HR). In response to the administration of metoprolol, statistically significant changes were observed in LVdp/dt, $\frac{LVdp/dt}{IP}$, CI, and HR, all of which decreased. In response to oxprenolol, a statistically significant decrease occurred in LVEDP, LVdp/dt, $\frac{LVdp/dt}{IP}$, HR, and $\Delta P/\Delta V$.

The following statistically significant differences were observed between the two drugs: LVEDP decreased with oxprenolol and increased with metoprolol; LVdp/dt and HR were less influenced by oxprenolol than by metoprolol; compliance was improved after oxprenolol, whereas it was practically unchanged after metoprolol. Figure 1 shows the influence of the drugs upon ventricular function as revealed by the relationship between left-ventricular end-diastolic pressure and left-ventricular systolic work index (LVSWI). Following metoprolol, ventricular function appeared to deteriorate in four patients, to improve slightly in one, and to shift along the same curve in the others. Following oxprenolol, ventricular function was favourably affected in four patients, remained practically unchanged in six, and showed a slight deterioration in one.

---

* Cattedra II Malattie Cardiovascolari, Università di Roma, Rome, Italy.

Table 1. Haemodynamic data at rest before (b) and after (a) metoprolol.

Case No.	Age	HR (beats/min.)	LVSP (mm. Hg)	LVEDP (mm. Hg)	AP (mm. Hg)	MAP (mm. Hg)	LVdp/dt (mm. Hg/sec.)	LVdp/dt / IP	CI (litres/min./m.²)	SVI (ml./beat/m.²)	LVSWI (g.m./beat/m.²)	ΔP/ΔV	P×R	SAR (dynes sec. cm.⁻⁵)	Diagnosis
1 b	32	95	116	11	116/66	90	2,678	49	2.2	23	25	0.23	11,020	1,557	Two-vessel disease
a		82	101	12	101/62	79	1,510	30	2.5	30	27	0.17	8,282	1,209	
2 b	49	62	135	13	135/56	82	2,010	47	1.6	25	23	0.25	8,370	2,184	Three-vessel disease + myocardial infarction
a		54	102	7	102/54	75	1,206	26	0.9	18	17	0.21	5,508	3,330	
3 b	54	90	150	9	150/79	112	1,983	28	1.6	18	25	0.20	13,500	2,925	Two-vessel disease + myocardial infarction
a		76	163	20	163/89	119	1,530	22	1.1	15	20	0.44	12,388	4,403	
4 b	42	80	116	9	116/68	90	2,235	38	1.9	24	26	0.21	9,280	2,140	One-vessel disease
a		68	122	10	122/69	93	1,440	24	1.7	25	28	0.20	8,296	2,382	
5 b	51	100	172	28	172/102	134	3,408	46	3.4	34	49	0.31	17,200	1,683	Three-vessel disease
a		80	165	28	165/100	127	2,414	33	2.6	32	43	0.31	13,200	2,088	
6 b	49	85	167	7	167/76	108	3,059	44	2.1	25	34	0.16	14,195	2,397	One-vessel disease
a		70	142	11	142/76	101	1,690	26	2	29	35	0.18	9,940	2,402	
7 b	46	66	115	14	115/56	80	1,189	28	1	14	12	0.42	7,590	3,805	Three-vessel disease + myocardial infarction
a		66	125	13	125/68	92	902	16	0.9	13	14	0.33	8,250	4,901	
8 b	61	64	195	17	195/73	130	2,210	39	2.1	33	50	0.27	12,480	2,623	Three-vessel disease
a		62	220	26	220/88	135	2,036	33	1.3	20	30	0.46	13,640	4,495	
9 b	53	102	141	28	141/85	108	2,287	36	1.4	13	14	0.78	14,382	3,596	Three-vessel disease + myocardial infarction
a		96	135	26	135/87	100	1,872	30	1.1	12	13	0.71	12,960	4,359	
10 b	46	72	100	26	100/69	85	1,524	35	1.3	19	15	0.57	7,200	2,695	Two-vessel disease + myocardial infarction
a		72	108	32	108/70	90	1,270	33	1.7	23	18	0.56	7,776	2,305	
11 b	50	82	114	10	104/64	86	1,467	27	3.2	39	40	0.09	9,348	1,145	Three-vessel disease
a		72	108	10	108/68	84	1,110	19	2.5	34	34	0.09	7,776	1,472	

LVSP = left-ventricular systolic pressure
LVEDP = left-ventricular end-diastolic pressure
AP = systolic and diastolic aortic pressure
MAP = mean aortic pressure
IP = isometric pressure
CI = cardiac index
SVI = stroke volume index
LVSWI = left-ventricular stroke work index
ΔP/ΔV = compliance
P×R = pressure-rate product, i.e. product of blood pressure times heart rate
SAR = systemic arterial resistance

Table 2. Haemodynamic data at rest before (b) and after (a) oxprenolol.

Case No.	Age	HR (beats/ min.)	LVSP (mm. Hg)	LVEDP (mm. Hg)	AP (mm. Hg)	MAP (mm. Hg)	LVdp/dt (mm. Hg/sec.)	LVdp/dt / IP	CI (litres/ min./ m.²)	SVI (ml./ beat/ m.²)	LVSWI (g.m./ beat/ m.²)	$\Delta P/\Delta V$	P×R	SAR (dynes sec. cm.$^{-5}$)	Diagnosis
1 b	52	90	124	28	124/78	$\overline{98}$	1,275	25	2.6	29	28	0.22	11,160	1,483	Three-vessel disease + myocardial infarction
a		90	118	24	118/79	$\overline{96}$	1,225	22	2.4	27	26	0.14	10,620	1,540	
2 b	60	68	200	14	200/93	$\overline{138}$	1,887	24	1.5	22	37	0.24	13,600	3,676	Three-vessel disease + myocardial infarction
a		62	193	12	193/87	$\overline{131}$	1,966	26	1.2	19	31	0.26	11,966	4,362	
3 b	54	72	146	8	146/76	$\overline{105}$	2,226	33	1	14	18	0.32	10,512	4,662	Three-vessel disease + myocardial infarction
a		72	140	7	140/80	$\overline{105}$	1,908	26	1.1	15	20	0.25	10,080	4,238	
4 b	61	62	128	7	128/50	$\overline{82}$	1,520	35	1.5	24	24	0.14	7,936	2,427	Two-vessel disease
a		60	136	9	136/61	$\overline{89}$	1,064	20	1.3	21	23	0.16	8,160	3,119	
5 b	41	76	116	17	116/63	$\overline{91}$	1,936	42	1.8	23	23	0.34	8,816	2,331	Two-vessel disease + myocardial infarction
a		76	120	13	120/69	$\overline{93}$	1,203	21	1.5	20	22	0.28	9,120	2,752	
6 b	44	95	153	7	153/79	$\overline{107}$	3,286	45	1.2	12	16	0.33	14,535	4,319	Two-vessel disease
a		82	158	6	158/79	$\overline{110}$	2,226	30	1	11	16	0.30	12,956	5,635	
7 b	40	80	123	12	123/78	$\overline{107}$	2,015	30	1.5	19	24	0.26	9,840	3,167	Three-vessel disease + myocardial infarction
a		76	128	9	122/78	$\overline{101}$	1,937	28	1.4	18	22	0.18	9,728	3,233	
8 b	46	74	127	15	127/62	$\overline{91}$	1,834	39	2.4	32	33	0.21	9,398	1,731	Two-vessel disease + myocardial infarction
a		68	123	12	123/67	$\overline{93}$	1,808	33	1.7	25	27	0.18	8,364	2,477	
9 b	46	104	131	38	131/78	$\overline{101}$	2,119	53	2.6	24	20	0.72	13,624	1,978	Three-vessel disease + myocardial infarction
a		94	115	30	115/71	$\overline{91}$	1,847	45	2.6	27	22	0.48	10,810	1,782	
10 b	62	60	193	17	193/67	$\overline{100}$	2,002	40	2.3	39	49	0.16	11,580	2,565	Two-vessel disease
a		57	172	10	172/59	$\overline{97}$	1,668	34	2.2	39	46	0.14	9,804	2,438	
11 b	52	82	186	16	186/96	$\overline{133}$	1,334	16	2.0	26	41	0.24	12,252	3,054	Two-vessel disease
a		80	173	9	177/85	$\overline{125}$	1,196	16	2.3	29	46	0.06	13,840	2,643	

Listed in Table 4 are the individual data – for heart rate, left-ventricular systolic and end-diastolic pressures, and pressure-rate product (i.e. product of blood pressure times heart rate) – recorded during exercise before and after beta-blockade. These variables were measured at the time of onset of anginal pain or, in the absence of such pain, just before termination of exercise.

In Table 5 the findings are expressed as mean changes from pre-drug control values. After both drugs, exercise was performed at a significantly lower heart rate and pressure-rate product and – though only after oxprenolol – also at a significantly lower left-ventricular systolic pressure than before treatment. Between the responses to the two drugs, however, there were no statistically significant differences.

To sum up, it appeared that, in terms of the effect upon overall cardiac performance, more favourable results were obtained with oxprenolol than with metoprolol.

Table 3. Mean changes ($\pm$ S.E.M.) in haemodynamic variables at rest as compared with pre-drug values.

	Metoprolol	Oxprenolol	Comparison between drugs
LVEDP (mm.Hg)	+ 2.1 ($\pm$ 1.5)	– 3.5** ($\pm$ 0.9)	$P < 0.01$
LVdp/dt (mm.Hg/sec.)	–643*** ($\pm$122)	–308* ($\pm$102)	$P < 0.05$
LVdp/dt / IP	– 11.4*** ($\pm$ 1.9)	– 7.4** ($\pm$ 2.1)	n.s.
CI (litres/min./m.^{2})	– 0.32* ($\pm$ 0.13)	– 0.15 ($\pm$ 0.08)	n.s.
SVI (ml./beat/m.^{2})	– 1.4 ($\pm$ 1.7)	– 1.2 ($\pm$ 0.9)	n.s.
HR (beats/min.)	– 9.1** ($\pm$ 2.0)	– 4.2** ($\pm$ 1.3)	$P < 0.05$
$\Delta P/\Delta V$	+ 0.02 ($\pm$ 0.03)	– 0.07* ($\pm$ 0.02)	$P < 0.05$

* $P < 0.05$ ** $P < 0.01$ *** $P < 0.001$

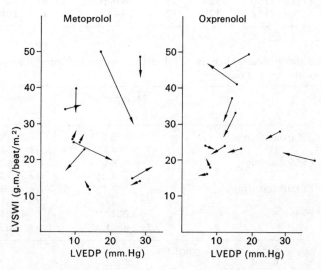

Fig. 1. Relationship between left-ventricular end-diastolic pressure (LVEDP) and left-ventricular systolic work index (LVSWI) before and after beta-blockade. Each arrow represents one patient, the arrow-head indicating the value recorded after administration of the beta-blocker.

The behaviour of left-ventricular end-diastolic pressure seems to be of particular importance, since this variable is closely related to ventricular function and compliance. In response to oxprenolol, left-ventricular end-diastolic pressure decreased in all but one patient, whereas in response to metoprolol it increased in six patients, showed no change in two, and decreased slightly in only three. It is conceivable that intrinsic sympathomimetic activity may be responsible for the differing circulatory effects of the two drugs under study. The fact that oxprenolol possesses such activity means

Table 4. Exercise data before (b) and after (a) beta-blockade.

Metoprolol					Oxprenolol				
Case No.	HR (beats/min.)	LVSP (mm. Hg)	LVEDP (mm. Hg)	P×R	Case No.	HR (beats/min.)	LVSP (mm. Hg)	LVEDP (mm. Hg)	P×R
1 b	145	141	33	20,445	2 b	96	186	25	17,856
a	132	123	36	16,236	a	92	178	27	16,376
2 b	124	125	31	15,500	3 b	128	169	34	21,632
a	90	121	14	10,890	a	108	167	20	18,036
3 b	114	190	27	21,660	4 b	118	162	29	19,116
a	108	170	29	18,360	a	106	153	25	16,218
4 b	148	134	33	19,832	5 b	122	161	31	19,642
a	120	129	32	15,480	a	118	144	28	16,992
6 b	130	169	10	21,970	6 b	140	223	15	31,220
a	106	183	17	19,398	a	106	209	18	22,154
7 b	114	136	36	15,504	7 b	126	160	30	20,160
a	98	138	33	13,524	a	120	162	31	19,440
8 b	90	226	34	20,340	8 b	118	156	22	18,408
a	98	192	30	18,816	a	106	139	22	14,734
9 b	180	157	30	28,260	9 b	156	151	46	23,556
a	144	156	34	22,464	a	104	130	34	13,520
10 b	124	144	32	17,856	10 b	110	200	28	22,000
a	116	130	31	15,080	a	95	193	24	18,335
11 b	150	126	23	18,900	11 b	130	187	29	24,310
a	112	120	22	13,440	a	106	180	29	19,080

Table 5. Mean changes (±S.E.M.) in haemodynamic variables during exercise as compared with pre-drug values.

	Metoprolol	Oxprenolol	Comparison between drugs
HR (beats/min.)	− 19.5** (± 4.8)	− 18.3** (± 4.7)	n.s.
LVSP (mm.Hg)	− 8.6 (± 4.2)	− 10.0** (± 2.3)	n.s.
LVEDP (mm.Hg)	− 1.1 (± 2.0)	− 3.1 (± 1.8)	n.s.
P×R	−3,657*** (± 459)	−4,301** (± 962)	n.s.

** $P < 0.01$ *** $P < 0.001$

that its negative inotropic effect is less marked and that compliance is improved (lower left-ventricular end-diastolic pressure with little change in stroke volume).

The role played by cardioselectivity, on the other hand, is less clearly apparent. Since of the two drugs only metoprolol is cardioselective, the more favourable haemodynamic changes obtained with oxprenolol cannot be ascribed to a difference in selectivity, particularly as the latter property could be expected to prove advantageous insofar as it helps to limit undesirable peripheral vascular effects.

It must be emphasised that this study involved an acute experiment, and we are not necessarily entitled to extrapolate the results to the situation of prolonged oral therapy. However, our data do seem to indicate that the choice of beta-blocker may be of importance in the therapeutic approach to patients with coronary artery disease in whom it is desirable that an already compromised cardiac performance should not be further impaired by pharmacological intervention.

References

1 AHLMARK, G., SAETRE, H.: Long-term treatment with beta-blockers after myocardial infarction. Europ. J. clin. Pharmacol. *10*, 77 (1976)
2 AHLMARK, G., SAETRE, H., KORSGREN, M.: Reduction of sudden deaths after myocardial infarction. Lancet *1974/II*, 1563; corresp.
3 GREEN, K. G., et al.: Improvement in prognosis of myocardial infarction by long-term beta-adrenoreceptor blockade using practolol. A multicentre international study. Brit. med. J. *1975/III*, 735
4 GREENBLATT, D. J., KOCH-WESER, J.: Adverse reactions to propranolol in hospitalized medical patients: a report from the Boston Collaborative Drug Surveillance Program. Amer. Heart J. *86*, 478 (1973)
5 HAMER, J., SOWTON, E.: Cardiac output after beta-adrenergic blockade in ischaemic heart disease. Brit. Heart J. *27*, 892 (1965)
6 HELFANT, R. H., HERMAN, M., GORLIN, R.: Abnormalities of left ventricular contraction induced by beta-adrenergic blockade. Circulation *43*, 641 (1971)
7 LAMBERT, D. M. D.: Hypertension and myocardial infarction. Brit. med. J. *1974/III*, 685; corresp.
8 PARKER, J. O., WEST, R. O., DI GIORGI, S.: Hemodynamic effects of propranolol in coronary artery disease. Amer. J. Cardiol. *21*, 11 (1968)
9 WILHELMSSON, C., VEDIN, J. A., WILHELMSEN, L., TIBBLIN, G., WERKÖ, L.: Reduction of sudden deaths after myocardial infarction by treatment with alprenolol. Preliminary results. Lancet *1974/II*, 1157

Influence of long-acting nitrates and oxprenolol on rehabilitation after acute myocardial infarction

by P. Rossi*, A. Tamitz**, A. Giordano**, and G. Minuco**

Social, psychological, and economic pressures compel patients who survive an acute myocardial infarction to resume a normal life as quickly as possible.
The success of cardiac rehabilitation is documented by the fact that a large proportion of patients actively participating in medically supervised training programmes are able to return to normal working conditions. This success, however, is limited in practice owing to the high spontaneous drop-out rate among participants[3].
Recent findings have shown that, in patients recovering from an acute myocardial infarction, mortality can be reduced by treatment with propranolol[9], alprenolol[22], or practolol[11].
Theoretically, patients receiving beta-blockers might be expected to prove more prone to cardiac failure, which in turn would make their rehabilitation more difficult. In the study to be described here, we therefore attempted to assess the possible influence of a beta-blocker and of long-acting nitrates on rehabilitation in patients who had survived an acute myocardial infarction.

Patients studied

The study was performed on 66 patients (60 men and six women) whose ages ranged from 51 to 59 years (median age: 55 years). All of them were patients who, 15 to 20 days after having suffered an acute infarction, had been admitted to the Rehabilitation Centre in Veruno, where they were then subjected to an endurance training programme.
The criteria adopted for the diagnosis of acute transmural infarctions were as follows: firstly, typical symptoms; secondly, unequivocal electrocardiographic changes; and, thirdly, changes in serum enzymes (glutamic oxalacetic transaminase, lactic dehydrogenase, and creatine phosphokinase). In 33 of the 66 patients the site of the infarction was anterior, in 29 inferior, in two postero-lateral, and in two postero-inferior.
The patients were randomly assigned to one of the following three medication groups:

Group 1 (18 patients): 40 mg. oxprenolol (®Trasicor) t.i.d.
Group 2 (23 patients): 7 mg. isosorbide dinitrate plus 25 mg. pentaerythrityl tetranitrate (®Stenodilate) t.i.d.
Group 3 (25 patients): 1 placebo tablet t.i.d.

All three groups were matched with respect to age, sex, heart volume, site of myocardial infarction, and reasons for stopping the basal exercise test.
The patients had been informed of the purpose of the study and had agreed to participate in it.

---

* Divisione di Cardiologia, Ospedale Maggiore della Carità, Novara, Italy.
** Fondazione Clinica del Lavoro, Centro Medico di Riabilitazione, Veruno (Novara), Italy.

Methodology

The patients were made to perform three exercise tests (E.T.s): the first (basal E.T.) during the fourth week after the occurrence of the acute infarction; the second (E.T. under medication) after three days of treatment with oxprenolol, nitrates, or placebo; and the third (E.T. under medication plus training) after a two-month training period during which the medication had been continued.

Prior to commencing the E.T., each patient had to remain in the lying position for at least ten minutes, while self-adhesive E.C.G. electrodes were placed in the deltoid region, in the groin, and in the best positions for thoracic leads (i.e. V_1, V_4, and V_6); in addition, a blood pressure cuff was attached round the upper arm. The E.C.G. from the three peripheral and from the three precordial leads, as well as the heart rate, were recorded continuously, and the blood pressure measured regularly once a minute, throughout the following successive periods: patient supine at rest for three minutes; patient sitting on bicycle ergometer; patient exercising; patient supine at rest for at least ten minutes. The E.C.G. tracings were recorded on a continuous paper strip using a six-channel apparatus (Mingograph 62-Elema Schonander Siemens).

The patients performed their exercise on a bicycle ergometer equipped with mechanical brakes. The work load was increased stepwise from 25 to 50, 75, and 100 watts, the duration of exercise at each level being six minutes so as to ensure the attainment of a steady state. The exercise was continued until the onset of one of the following signs or symptoms: severe anginal pain, fatigue, dyspnoea, pain in the legs, pronounced and rapidly increasing changes in the ST segment, ectopic beats of high and increasing frequency (especially if multifocal, occurring in series, or associated with the R-on-T phenomenon), tachyarrhythmia, prolongation of the PR interval, or increasing broadening of the QRS complex.

For the purposes of statistical analysis, non-parametric tests were employed.

Description of training programme

During the training programme, each patient was advised not to allow his heart rate to rise beyond 80% of the maximum rate recorded in the course of the E.T. The patients began the programme at a low level of activity, kept daily records of their symptoms and performance, and did not progress to the next level of activity till instructed to do so. Such instructions were given only after the physician, having carried out a clinical examination and monitored the patient's heart rate and blood pressure during exercise at the previous level of activity, had convinced himself that there was no evidence of heart failure or any other complications. The training programme involved varying levels of gymnastic, walking, and bicycling activity, which were individually adapted for patients of both sexes. Each class was placed under the direction of a physical trainer, and a physician was always available to provide clinical advice. For use in an emergency, an appropriate selection of drugs, an electrocardiograph, and a direct-current defibrillator were kept in the training room.

Results

The reasons for stopping the E.T.s in patients belonging to each of the three medication groups are listed in Table 1. During the first E.T., the appearance of severe

Table 1. Reasons for stopping exercise tests (E.T.s) in each of the three medication groups.

Groups	(1 oxprenolol)			(2 nitrates)			(3 placebo)		
	1st	2nd	3rd E.T.	1st	2nd	3rd E.T.	1st	2nd	3rd E.T.
Reasons:									
Plain fatigue	11	15	13	11	11	19	17	17	20
	61	*83*	*72*	*48*	*48*	*83*	*68*	*68*	*80*
Severe E.C.G. changes	7	3	4	9	9	2	7	7	5
	39	*17*	*22*	*39*	*39*	*9*	*28*	*28*	*20*
Anginal pain	3	0	1	3	3	2	3	3	1
	17	*0*	*5*	*13*	*13*	*9*	*9*	*9*	*3*

The upper rows of figures indicate numbers of patients and the figures below in italics the percentages. In some instances these percentages add up to over 100% because there was more than one reason for stopping the test.

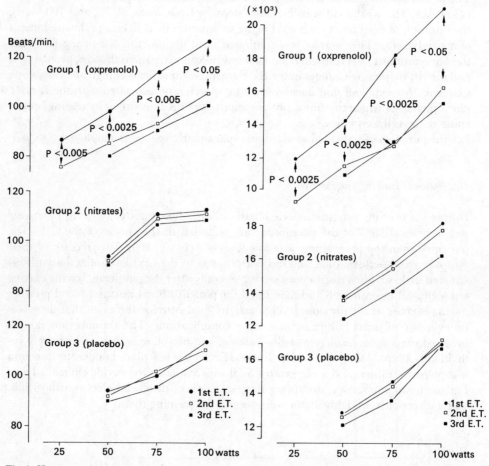

Fig. 1. Heart rate increments (mean values) at different work loads in the three medication groups during the three exercise tests (E.T.s).

Fig. 2. Pressure-rate product (mean values) at different work loads in the three medication groups during the three exercise tests (E.T.s).

E.C.G. changes (pronounced ST deviation) compelled the investigator to break off the exercise in 28% of the placebo-treated patients and in 39% of the patients both in the oxprenolol group and in the nitrate group.

During the second E.T., the emergence of severe E.C.G. changes necessitated the cessation of exercise in the oxprenolol group less frequently (17%, $P<0.05$) than during the first E.T.

In all three groups, plain fatigue had been the commonest reason for breaking off the exercise during the first E.T. During the second E.T., however, the fact that the oxprenolol group showed an enhanced maximum performance meant that in this group plain fatigue was slightly more often responsible for discontinuation of the exercise than in the other two groups.

During the second E.T., it was only in the oxprenolol group that both the maximum heart rate (Figure 1) and pressure-rate product, i.e. the product of blood pressure times heart rate (Figure 2), were found to be significantly lower than during the first E.T., the respective P values at the successively increasing work loads of 25, 50, 75, and 100 watts being <0.005, <0.0025, <0.005, and <0.05 for heart rate and <0.0025, <0.0025, <0.0025, and <0.05 for pressure-rate product. In the other two groups there were no differences in these respects between the first and the second E.T.

Maximum performance increased significantly in all three groups after completion of the training period (Figure 3): in the oxprenolol group it increased from 83.3 to 106.9 watts ($P<0.005$) at the second E.T. and to 125.0 watts ($P<0.005$) at the third E.T.; in the other two groups, by contrast, the maximum performance showed no difference between the first and second E.T. and underwent a significant increase ($P<0.005$) only in the third E.T.

From the maximum performance recorded at the third as compared with the first E.T. it is apparent that performance increased by 42 watts in the oxprenolol group,

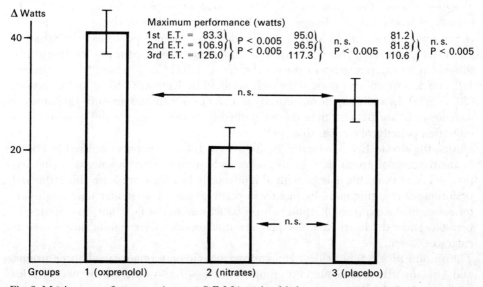

Fig. 3. Maximum performances (mean ± S.E.M.) at the third, as compared with the first, exercise test (E.T.).

Table 2. Mean heart volume values in the three medication groups (expressed in cubic centimetres per square metre of body surface area) before and after the training programme.

Groups	Before	After	Significance
1 (oxprenolol)	383.5	365.9	n.s.
2 (nitrates)	363.0	343.4	n.s.
3 (placebo)	406.3	348.0	$P < 0.05$

as against 22 watts and 29 watts, respectively, in the nitrate group and the placebo group (Figure 3).

Indicated in Table 2 are the heart volume values measured before, and after completion of, the training programme: a trend towards a reduction in heart volume was apparent in the oxprenolol and nitrate groups, whereas in the placebo group the reduction was statistically significant ($P<0.05$).

Discussion

Dynamic conditioning exercise and endurance training programmes are capable of increasing functional aerobic capacity (maximum oxygen uptake) both in sedentary healthy subjects[1,7,12-14,19] and in ambulant patients with cardiac diseases[4-6,10,15,18,20,21].
The degree of functional improvement achieved depends on how interested and well-motivated the patient is and upon the extent to which he is willing to adhere to the training programme for months or even years.

Although the physical capacity of our patients, who were studied at an earlier stage after their acute myocardial infarction than those reported upon by other authors[2,8,16-18,20], was probably not inferior to that of the latter patients, direct comparisons between the two are impossible.

The percentages of our patients who had to break off exercise during the first E.T. because of severe E.C.G. changes (i.e. 39%, 39%, and 28% in Groups 1, 2, and 3, respectively) tally well with those recorded in the studies published by SANNE[20], KENTALA[17], and IBSEN et al.[16]. In our series, however, anginal pain was less frequently the reason for stopping exercise during the first E.T. (17%, 13%, and 9% in Groups 1, 2, and 3, respectively) than in the series studied by KENTALA (40%) and by SANNE (33–38.5%). In addition, the maximum values for heart rate and pressure-rate product were lower in our patients than those obtained in healthy subjects and in other post-infarction patients of comparable age[16].

During the second E.T., anginal pain and severe E.C.G. changes occurred less often in the oxprenolol group than in the other two groups. The patients receiving oxprenolol were thus able to cope with a significantly heavier work load, this improved performance resulting more frequently in plain fatigue. The greater maximum performance in the oxprenolol group was probably due to the fact that, in response to beta-blockade, the heart rate and pressure-rate product were significantly lower at each work load.

The training programme achieved an improvement in performance in all three groups, and – in the nitrate and placebo groups as well – it also reduced the necessity for stopping exercise because of severe E.C.G. changes or angina pectoris. A decrease in heart volume in response to training was observed in each of the three groups – this observation being of particular importance in the case of the oxprenolol group since it suggests that no latent heart failure was present.

In conclusion, the findings obtained in our study serve to confirm that cardiac rehabilitation in the form of a medically supervised programme of physical training can yield successful results. All three groups improved their maximum performance, and – although in this respect the oxprenolol and nitrate groups did not differ significantly from the placebo group – the better performance of the oxprenolol group during the second E.T. does suggest that beta-blockade can safely be resorted to in patients recovering from an acute myocardial infarction and that it is of value in cardiac rehabilitation.

References

1 ÅSTRAND, I.: Aerobic work capacity in men and women with special reference to age. Acta physiol. scand. *49*, Suppl. 169 (1960)
2 ATTERHÖG, J.-H., EKELUND, L.-G., KAIJSER, L.: Electrocardiographic abnormalities during exercise 3 weeks to 18 months after anterior myocardial infarction. Brit. Heart J. *33*, 871 (1971)
3 BRUCE, E.H., FREDERICK, R., BRUCE, R.A., FISHER, L.D.: Comparison of active participants and dropouts in CAPRI cardiopulmonary rehabilitation programs. Amer. J. Cardiol. *37*, 53 (1976)
4 CLAUSEN, J.P., LARSEN, O.A., TRAP-JENSEN, J.: Physical training in the management of coronary artery disease. Circulation *40*, 143 (1969)
5 COUNCIL ON REHABILITATION, INTERNATIONAL SOCIETY OF CARDIOLOGY: Myocardial infarction, how to prevent, how to rehabilitate (Council on Rehabilitation, International Society of Cardiology, 1973)
6 DETRY, J.-M., ROUSSEAU, M., VANDENBROUCKE, G., KUSUMI, F., BRASSEUR, L.A., BRUCE, R.A.: Increased arteriovenous oxygen difference after physical training in coronary heart disease. Circulation *44*, 109 (1971)
7 EKBLOM, B., ÅSTRAND, P.-O., SALTIN, B., STERNBERG, J., WALLSTRÖM, B.: Effect of training on circulatory response to exercise. J. appl. Physiol. *24*, 518 (1968)
8 ERICSSON, M., GRANATH, A., OHLSÉN, P., SÖDERMARK, T., VOLPE, U.: Arrhythmias and symptoms during treadmill testing three weeks after myocardial infarction in 100 patients. Brit. Heart J. *35*, 787 (1973)
9 FOX, K.M., CHOPRA, M.P., PORTAL, R.W., ABER, C.P.: Long-term beta blockade: possible protection from myocardial infarction. Brit. med. J. *1975/I*, 117
10 FRICK, M.H., KATILA, M.: Hemodynamic consequences of physical training after myocardial infarction. Circulation *37*, 192 (1968)
11 GREEN, K.G., et al.: Improvement in prognosis of myocardial infarction by long-term beta-adrenoreceptor blockade using practolol. A multicentre international study. Brit. med. J. *1975/III*, 735
12 GRIMBY, G., BJURE, J., AURELL, M., EKSTRÖM-JODAL, D., TIBBLIN, G., WILHELMSEN, L.: Work capacity and physiologic responses to work – men born in 1913. Amer. J. Cardiol. *30*, 37 (1972)
13 HANSON, J.S., TABAKIN, B.S., LEVY, A.M., NEDDE, W.: Long-term physical training and cardiovascular dynamics in middle-aged men. Circulation *38*, 783 (1968)
14 HARTLEY, L.H., GRIMBY, G., KILBOM, A., NILSON, N.J., ÅSTRAND, I., BJURE, J., EKBLOM, B., SALTIN, B.: Physical training in sedentary middle-aged and older men – cardiac output and gas exchange at submaximal and maximal exercise. Scand. J. clin. Lab. Invest. *24*, 335 (1969)
15 HELLERSTEIN, H.K., FORD, A.B.: Rehabilitation of the cardiac patient. J. Amer. med. Ass. *164*, 225 (1957)
16 IBSEN, H., KJØLLER, E., STYPEREK, J., PEDERSEN, A.: Routine exercise ECG three weeks after acute myocardial infarction. Acta med. scand. *198*, 463 (1975)
17 KENTALA, E.: Physical fitness and feasibility of physical rehabilitation after myocardial infarction in men of working age. Ann. clin. Res. *4*, Suppl. 9 (1972)
18 ROUSSEAU, M., BRASSEUR, L.A., DETRY, J.-M.: Hemodynamic determinants of maximal oxygen intake in patients with healed myocardial infarction: influence of physical training. Circulation *48*, 943 (1973)

19 SALTIN, B., BLOMQVIST, G., MITCHELL, J.H., JOHNSON, R.L., Jr., WILDENTHAL, K., CHAPMAN, C.B.: Response to exercise after bed rest and after training. Circulation *38*, Suppl.VII (1968)
20 SANNE, H.: Exercise tolerance and physical training of non-selected patients after myocardial infarction. Acta med. scand. *194*, Suppl. 551 (1973)
21 VARNAUSKAS, E., BERGMAN, H., HOUK, P., BJÖRNTORP, P.: Haemodynamic effects in physical training in coronary patients. Lancet *1966/II*, 8
22 WILHELMSSON, C., VEDIN, J.A., WILHELMSEN, L., TIBBLIN, G., WERKÖ, L.: Reduction of sudden deaths after myocardial infarction by treatment with alprenolol. Preliminary results. Lancet *1974/II*, 1157

The release and uptake of catecholamines by the myocardium in patients with coronary artery disease

by W. KÜBLER and W. MÄURER*

In patients with coronary artery disease, it is believed that changes in sympatho-adrenergic activity may play a role in the development and manifestation of the disease and that such changes may serve as a mechanism to compensate for the impairment of myocardial function.

Sympatho-adrenergic activity can be assessed in these patients by determining their plasma catecholamine concentrations. Using the highly specific and extremely sensitive double-isotope method, it is possible to measure the levels not only of noradrenaline and adrenaline but (thanks to a modification which has been introduced by MÄURER) also of dopamine.

In the presence of coronary artery disease and impaired left-ventricular function, elevated noradrenaline and adrenaline levels are found in the plasma together with corresponding increases in the noradrenaline and adrenaline content of the erythrocytes. This finding – coupled with earlier observations of relevance in this connection – suggests the existence of an equilibrating system involving the permeation of catecholamines from the plasma into the erythrocytes. Under normal conditions, the noradrenaline and adrenaline content of the blood platelets far exceeds the levels measured in the plasma and the erythrocytes. In patients with coronary artery disease and impaired left-ventricular function the platelets have an even higher content of catecholamines, which corresponds to the latter's increased plasma concentrations; on the other hand, the dopamine levels in the plasma and erythrocytes of such patients show no significant changes. In coronary artery disease associated with left-ventricular failure an increase thus occurs not only in the plasma concentrations of noradrenaline and adrenaline but also in the content of these catecholamines in the blood cells, including especially the platelets.

The heart can be regarded as an endocrine organ synthesising noradrenaline. By simultaneously determining the coronary arterio-venous difference and the coronary blood flow with the aid of the highly sensitive argon method, it is possible to measure the uptake or release of catecholamines under various conditions.

In subjects at rest adrenaline, as well as noradrenaline and dopamine, are taken up by the heart from arterial blood. When the handgrip test is performed, however, i.e. when the muscles of the hand are isometrically tensed, the myocardium increases its uptake of adrenaline and dopamine while at the same time releasing noradrenaline.

In patients with a diminished coronary reserve the heart releases greater quantities of noradrenaline during the handgrip test. When this test is carried out by patients with severe obstructive lesions of the coronary arteries, a negative correlation can be observed between the release of noradrenaline from the myocardium and the extent to which coronary reserve is diminished.

---

* Abteilung Innere Medizin III (Kardiologie), Medizinische Universitätsklinik, Heidelberg, German Federal Republic.

The role played by the catecholamines in the pathogenesis and manifestation of coronary artery disease may conceivably be as follows: as a result of the increase in the work performed by the heart, the myocardium releases noradrenaline, which, by promoting local ischaemia, may possibly give rise to an acute infarction associated with disturbances in cardiac rhythm and leading to sudden death.

Clinical aspects of the treatment of cardiovascular diseases with beta-blocking drugs

by S. H. Taylor*

The epidemic increase in coronary heart disease has fortunately coincided with the advent of a "natural" pharmacological solution in the shape of the beta-adrenoceptor antagonists – a coincidence which is perhaps not entirely fortuitous. The dangers inherent in this disease are not solely due to the pathological state of the arteries supplying the heart; they are also closely associated with the electrical and mechanical demands made on the myocardium supplied by these diseased vessels. Such demands are conveyed predominantly through the sympathetic nervous control of the heart, the final effector pathway of which is the beta-adrenoceptor in the myocardium. It is therefore easy to understand the logic of employing drugs which – by attenuating sympathetic influences on the heart, whether these have their source in psychic stress or physical exertion – serve to avert the hazard of catecholamine stimulation of the heart afflicted by obstructive arterial disease.

The detailed pharmacological basis for the cardiovascular action of beta-adrenoceptor antagonists in patients at risk from coronary heart disease has been reviewed by Brunner[1]. The antihypertensive role of these agents[16] and their influence in preventing infarction in patients with angina pectoris[8] have also been described. Wilhelmsson et al.[15] have reported on the positive results of their prevention trials in patients with coronary heart disease. It might thus perhaps be assumed that there can be little else to say about the preventive value of treatment with beta-blocking drugs in patients at risk from coronary heart disease. But an essential prerequisite for the exploitation of such treatment by the practising physician is the rational prescription of these drugs in the various cardiovascular diseases in which they can be used. It is the aim of the following review to deal with this essentially practical problem.

Hypertension

Zanchetti[16] has elegantly described the possible mechanisms underlying the antihypertensive effects of beta-adrenoceptor antagonists in hypertensive patients. But there is still a relative paucity of information on the detailed prescribing aspects of these drugs, in connection with which a number of questions have to be answered:

1. What is the rapidity of onset and duration of action of a single oral dose?
2. What is the magnitude of the effects on blood pressure at rest and during exercise?
3. What are the immediate and long-term dose-response relationships of the various drugs available?
4. Do these drugs have a cumulative effect?
5. How do different beta-blocking drugs compare with one another with respect to their antihypertensive effectiveness?

---

* University Department of Cardiovascular Studies and Department of Medical Cardiology, The General Infirmary, Leeds, England.

During the past few years a programme of study has been initiated in our laboratory in an attempt to answer these vital prescription problems. A range of beta-adrenoceptor antagonists with differing ancillary pharmacological properties (Table 1) have been studied both acutely[9] and during long-term treatment[2] in a series of double-blind randomised crossover factorial comparisons. To simplify matters, the results obtained will be illustrated by reference to one of these drugs, i.e. oxprenolol (®Trasicor).

Sympatho-adrenal influences are a major determinant of the rate of rise in pressure in the left ventricle (dp/dt max.) and the peak systolic pressure attained. Measurement of the latter variable thus provides a useful and informative non-invasive method of assessing the degree of sympathetic stimulation of the heart. Moreover, as one of

Table 1. Potency, ancillary pharmacological properties, and clinical dose ranges of five beta-adrenoceptor antagonists.

Drug	Potency (ΔHR_{Ex})	Membrane-stabilising activity	Intrinsic sympatho-mimetic activity	Cardio-selectivity	Single-dose range (mg.)
Propranolol	1	+	0	0	40–320
Oxprenolol	1	+	+	0	40–320
Practolol	2/5	0	+	+	100–800
Metoprolol	1	0	0	+	50–400
Tolamolol*	1	0	±	+	50–400

* Possesses weak vasodilator activity

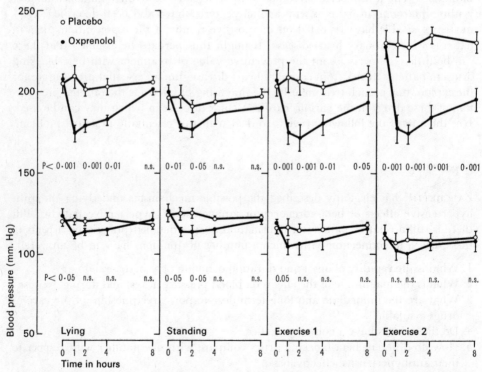

Fig. 1. Duration of antihypertensive activity after a single oral dose of oxprenolol (160 mg.) in patients with stable, uncomplicated essential hypertension. Data plotted as mean ± S.E.M. The significance of the differences relates to comparisons between placebo and oxprenolol.

the major principles governing the action of pharmacological antagonists is that their effects are proportional to the agonist stimulus applied, it is to be anticipated that the effects of the beta-adrenoceptor antagonists will be most apparent on the systolic blood pressure during exercise. Attention will therefore be particularly directed to this variable.

1. Rapidity of onset and duration of action of a single oral dose (Figure 1)
The pharmacodynamic effects of oxprenolol, especially on the systolic blood pressure during exercise, are clearly discernible within an hour of the drug's ingestion. These antihypertensive effects are maintained for up to four hours, after which they gradually decline.

2. Effects on blood pressure at rest and during exercise (Figures 1–3)
It is during exercise that all beta-adrenoceptor antagonists, including oxprenolol, produce their maximum antipressor activity on the systolic pressure.

3. Immediate and long-term dose-response relationships (Figures 2 and 3)
In comparison with placebo, oxprenolol exerts an immediate antihypertensive effect which produces a significant reduction in exercise systolic pressure related to the logarithm of the dose. The most clinically efficacious single oral antipressor dose is 80–160 mg. in subjects of 70–80 kg. body weight. There is a similar decremental relationship between the dose and the antihypertensive effect of the drug during long-term treatment. Under these conditions, the most effective dose appears to be in the range of 80–160 mg. twice daily.

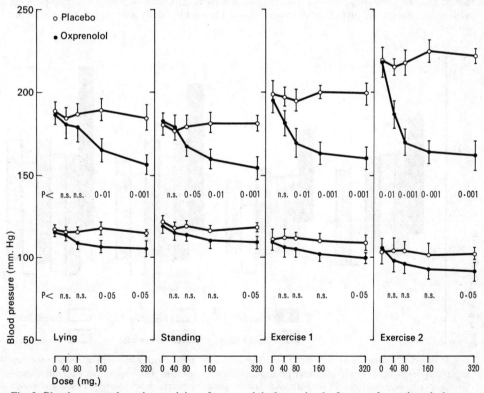

Fig. 2. Blood-pressure lowering activity of oxprenolol, determined after one hour, in relation to size of oral dose. Results obtained in patients with stable, uncomplicated essential hypertension.

4. Cumulative effect of continued treatment

The cumulative effect of beta-adrenoceptor antagonists may be due to two causes. Firstly, the drug or its active metabolites may accumulate in the tissues: oxprenolol

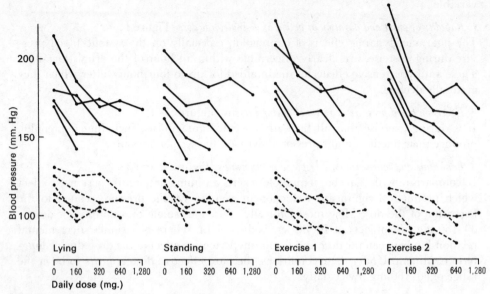

Fig. 3. Dose-response relationships during sustained treatment with oxprenolol in five patients with stable, uncomplicated essential hypertension. The divided daily dose was given at 12-hour intervals.

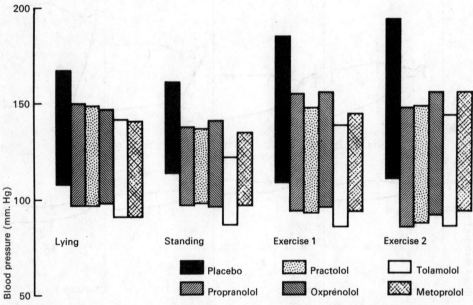

Fig. 4. Blood-pressure lowering activity during sustained treatment with beta-adrenoceptor antagonists possessing different ancillary pharmacological properties. Data from a double-blind randomised crossover study in 25 patients with stable, uncomplicated essential hypertension. Shown in the figure are the blood pressure values (determined one hour after an oral dose) at the end of eight weeks' treatment.

does not accumulate in the tissues and no active metabolites have yet been detected. Secondly, the long-term effects of the drug may incorporate influences due to persistent modulation of sympathetic activity in the hypertensive patient, a modulation possibly related to the chain of events precipitated by chronic reduction of renin and aldosterone secretion. Hypertensive patients on a constant oral dose of oxprenolol exhibit a maximum antihypertensive effect after between two and four weeks. From the prescribing standpoint, this means that hardly any additional antihypertensive effect can be expected after one month's treatment with a constant daily dose of the drug.

5. Comparative antihypertensive activity of different beta-blocking drugs (Figure 4)
All beta-adrenoceptor antagonists appear to have similar antipressor activity in hypertensive patients. This indicates that, whatever may be the value of their ancillary pharmacological properties from other points of view, the antihypertensive effect of these drugs is independent of such properties.

Angina pectoris

RIVIER[8] has presented a review of the cardiac risks attending patients with angina pectoris and has described how these may be attenuated with beta-blocking drugs. However, since relatively little information has been published on the practical problems of prescribing these drugs in such patients, questions similar to those posed with regard to hypertension also arise in this connection. In our laboratory we have attempted to answer them in a series of double-blind randomised crossover trials[14] using the same drugs as listed in Table 1. Here again, the results we obtained will be illustrated by reference to the example of oxprenolol[11].

1. Rapidity of onset and duration of action of a single oral dose (Figures 5 and 6)
Following a single oral dose of oxprenolol a significant increase in walking distance and an electrocardiographic improvement were observed in patients with exercise-induced angina pectoris; the response set in within an hour of the drug's ingestion and persisted for over eight hours.

2. Immediate and long-term dose-response relationships (Figures 7 and 8)
Both at the start of medication and during sustained treatment there was a significant relationship between anti-anginal activity (walking distance and ST depression in the E.C.G.) and the logarithm of the dose of oxprenolol.

3. Cumulative effect of continued treatment (Figure 9)
It is of considerable theoretical interest, and also of importance from the prescribing angle, to note that the dose-response curves plotted in the same patients at the start of medication and after sustained treatment exhibited a different quantitative relationship. The practical conclusion emerging from these findings is that the same dose of oxprenolol will exert significantly greater anti-anginal activity as part of continued treatment than when first given.

4. Comparative anti-anginal activity of different beta-blocking drugs (Figure 10)
All beta-adrenoceptor antagonists appear to exert similar anti-anginal activity in terms of increased exercise tolerance and improvement in E.C.G. evidence of myocardial ischaemia. Exercise-induced tachycardia and the systolic pressor response to exertion are attenuated to similar degrees by all these drugs. This implies that their anti-

anginal activity is a function solely of their ability to block cardiac beta-receptors and is independent of their ancillary pharmacological spectra.

Coronary heart disease

Beta-adrenoceptor antagonists reduce the area of myocardial infarction after coronary occlusion in the experimental animal[6]. In clinical practice it now seems clear

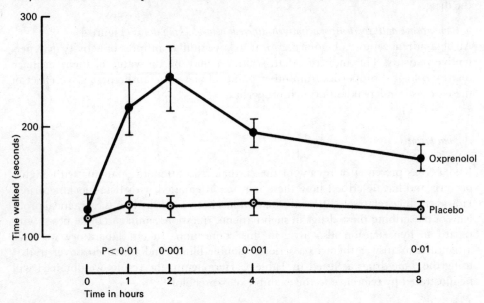

Fig. 5. Duration of action of a single oral dose of oxprenolol (160 mg.) in patients with stable, uncomplicated exercise-induced angina pectoris. Data plotted as mean ± S.E.M. The significance of the differences relates to comparisons between placebo and oxprenolol.

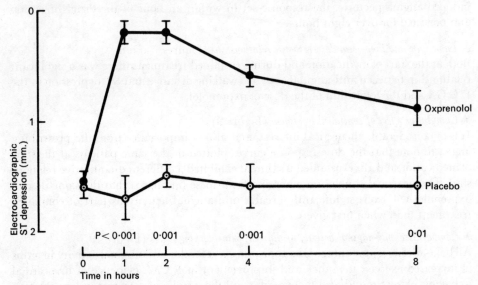

Fig. 6. Duration of effects on electrocardiogram of a single oral dose of oxprenolol (160 mg.) during walking in patients with stable, uncomplicated exercise-induced angina pectoris.

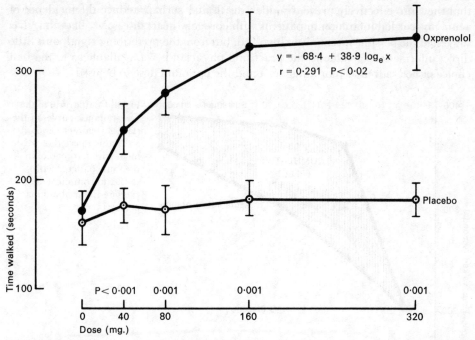

Fig. 7. Relationship between dose and increase in time walked before the onset of pain in patients with stable, uncomplicated exercise-induced angina pectoris.

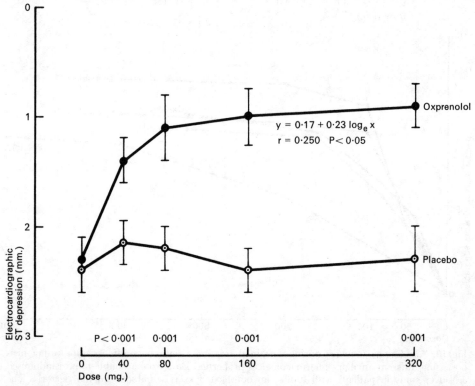

Fig. 8. Relationship between dose of oxprenolol and reduction in electrocardiographic ST depression in patients with exercise-induced angina pectoris.

that they also effectively prevent sudden death and perhaps reduce the incidence of acute myocardial infarction in patients with coronary heart disease[4,5]. BRUNNER[1] has suggested the possible mechanisms involved, but from the prescribing standpoint little direct information of value to the practising physician is yet available as to the best choice of beta-adrenoceptor antagonist and the optimum dose to be used.

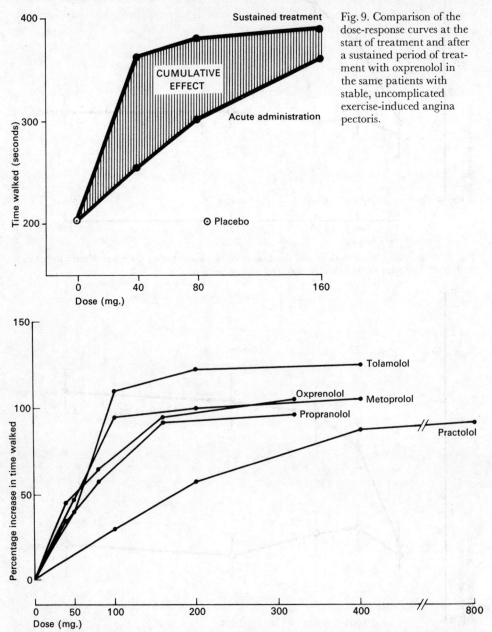

Fig. 9. Comparison of the dose-response curves at the start of treatment and after a sustained period of treatment with oxprenolol in the same patients with stable, uncomplicated exercise-induced angina pectoris.

Fig. 10. Anti-anginal activity during sustained treatment with beta-adrenoceptor antagonists possessing different ancillary pharmacological properties. Data from a double-blind randomised crossover study in patients with stable, uncomplicated exercise-induced angina pectoris. The results shown (recorded one hour after an oral dose) were obtained at the end of eight weeks' treatment on each drug.

Two aspects need to be considered in this context. The first is concerned with sympathetic stimulation of the heart provoked by exercise or by catecholamines released in response to psychic (mental) stress. The initial tachycardia of exercise is almost entirely related to vagal withdrawal, sympathetic stimulation only becoming predominant at heart rates in excess of 120 per minute; the influence of sympathetic drive increases exponentially with the heart rate. This contrasts with the cardiac stimulation and tachycardia induced by blood-borne catecholamines, in which the response appears to be linearly related to the amount infused. The second aspect relates to the ability of various beta-blocking drugs to counteract the tachycardia induced by circulating catecholamines. As antagonists of infused catecholamines, beta-adrenoceptor antagonists without cardioselective activity appear to have a much more potent action than those endowed with this selective property[12]. In this respect, oxprenolol is a far

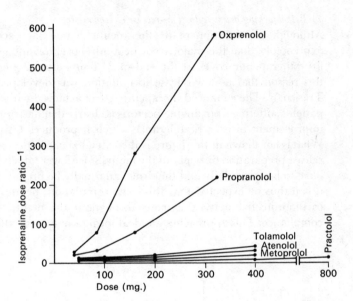

Fig. 11. Relationship between dose of beta-adrenoceptor antagonist and antagonism of isoprenaline-induced tachycardia recorded in normal subjects in a double-blind randomised trial. The isoprenaline dose ratio^{-1} is the dose of isoprenaline required to produce a standard increase in heart rate (20 beats/min.) in a person pre-treated with a given dose of beta-adrenoceptor antagonist, divided by the dose of isoprenaline required to produce the same increase when no pre-treatment has been administered.

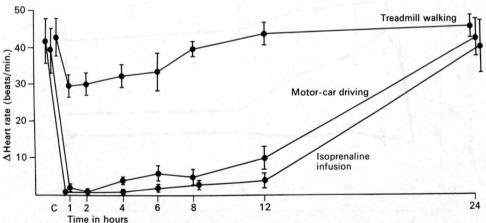

Fig. 12. Difference in the effects of oxprenolol (40 mg.) on the tachycardia associated with motor-car driving, isoprenaline infusion, and treadmill walking in six normal subjects. The control values (C) represent the average of the increases in heart rate measured on placebo. Data plotted as mean ± S.E.M.

more potent antagonist of infused isoprenaline and adrenaline than cardioselective drugs as a whole (Figure 11). Hence, even relatively small doses of oxprenolol are capable of effectively countering very large increases in circulating catecholamines. On the other hand, much larger doses of the drug are necessary to counter similar degrees of tachycardia due to exercise. As practising clinicians, we can therefore anticipate that small doses of oxprenolol – e.g. 40 mg. b.i.d. – may effectively combat the cardiac effects of psychic stress while causing comparatively little reduction in exercise tolerance[10] (Figure 12).

New advances in the use of beta-adrenoceptor antagonists in preventing the consequences of coronary heart disease

a) Beta-blocking drugs with slow-release characteristics

Although the duration of its therapeutic action far exceeds its plasma half-life, oxprenolol – like the majority of beta-adrenoceptor antagonists – affords relatively little therapeutic cover at the end of 24 hours after a single oral dose[12]. It was for this reason that a slow-release formulation was developed for oxprenolol (®Slow-Trasicor)[3]. The enhanced therapeutic effect achieved with slow-release oxprenolol in patients suffering from angina pectoris is clearly demonstrated by the more prolonged improvement in exercise tolerance which it produces[13], as illustrated in Figure 13. What is not shown in this figure is the extended cardioprotection offered by this slow-release preparation throughout the night. Since fewer tablets of slow-release oxprenolol have to be taken daily and the drug can usually be prescribed on a once-a-day basis, it can thus be expected that this new formulation, besides ensuring that the drug's cardioprotective activity extends throughout the night, will also improve patient compliance. This approaches an ideal form of treatment affording a cardioprotective

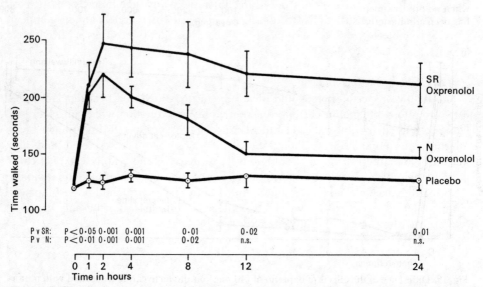

Fig. 13. Difference in the duration of action of a dose of 160 mg. oxprenolol in the normal formulation (N) and the same dose of the slow-release formulation (SR), determined by reference to the time walked by patients with angina pectoris.

action against exercise-induced stimulation during the day and countering possible psychic stimulation (as provoked by dreams) during the night.

b) Ancillary pharmacological properties

There is no doubt that the major therapeutic activity of the beta-adrenoceptor antagonists is directly related to their ability to counteract sympathetic stimulation at beta-adrenoceptor sites in the heart. The ancillary pharmacological properties exhibited by each of these drugs are nevertheless worthy of comment in terms of the cardioprotection they can offer.

Membrane-stabilising activity is a potentially important therapeutic property possessed by both propranolol and oxprenolol. This property is characterised by the capacity to reduce the electrical activity of the myocardial cell membrane, thus diminishing intrinsic excitability and decreasing sensitivity to catecholamine stimuli. In therapeutic terms it offers the potential advantage of decreasing the risk of serious cardiac dysrhythmias, but it involves the disadvantage of being associated with a reduction in the contractile activity of the myocardial cell. The doses usually employed in clinical practice, however, rarely result in tissue concentrations high enough to produce this effect.

Cardioselectivity is a property that affords no discernible cardioprotective advantage. In fact, it may even be a disadvantage: beta-blocking drugs displaying this property (practolol, atenolol, metoprolol, acebutolol) appear to be considerably less potent in antagonising the effects of circulating catecholamines on the heart than those that are devoid of cardioselectivity (propranolol and oxprenolol).

Intrinsic sympathomimetic activity, as exhibited by oxprenolol for example, is associated with the possession by the drug's molecule of partial agonist (stimulating) activity. It is a property about which there has been much debate concerning its clinical importance, and it may possibly be of greater therapeutic significance than was originally realised. Bradycardia and slowing of conduction in the atrioventricular node and conducting system of the heart are less marked in response to drugs with this property than to those without it. Drugs possessing intrinsic sympathomimetic activity give rise to less airway obstruction (GARRARD, unpublished observations) and also appear to be less liable to produce peripheral vascular changes[7] than drugs with no such activity.

c) The ideal beta-blocking drug

What are the properties that should ideally be incorporated in a beta-blocking drug if it is to afford maximum cardioprotection? This is a question which all clinicians who use these drugs must ask. The ideal drug should be a specific and potent beta-adrenoceptor antagonist. It should be rapidly absorbed and excreted and also "cleanly" metabolised, i.e. it should convert to inactive and readily excreted compounds by simple metabolic transformations. For maximum cardioprotection it should *not* be cardioselective, and for maximum therapeutic efficacy it should possess intrinsic stimulating activity. Oxprenolol and alprenolol (®Gubernal) are the only drugs in widespread clinical use that fulfil these criteria.

Summary

Beta-receptor antagonists have been shown to exhibit a cardioprotective action in the three major disease states of which sudden death and myocardial infarction are the commonest consequences. There is still a relative paucity of information on the prescribing aspects of these drugs, but it would appear that patients with hypertension or angina pectoris require 80–160 mg. oxprenolol, or an equivalent dose of an alternative drug, twice daily to achieve maximum benefit. To counter the effects of stress-induced catecholamine stimulation of the heart, the dose required is much smaller, namely 40 mg. oxprenolol twice daily or its equivalent. The advent of a slow-release formulation for oxprenolol offers significant advantages, not only because it makes for improved patient compliance, but also because it enables blood concentrations to be maintained which are sufficient to modulate the cardiac effects of exercise during the day and to suppress the catecholamine hazards of psychic stimulation at night.

Finally, it is argued that the only ancillary pharmacological property of value in a beta-adrenoceptor antagonist is probably intrinsic sympathomimetic activity, and the fact that oxprenolol possesses such activity, coupled with a "clean" metabolism, emphasises its pre-eminent role as the cardioprotective beta-blocking drug of choice.

References

1 BRUNNER, H.: The pharmacological basis for a cardioprotective action of beta-blockers. In Gross, F. (Editor): The cardioprotective action of beta-blockers, Int. Symp., Amsterdam 1976, p. 11 (Huber, Berne/Stuttgart/Vienna 1977)
2 DAVIDSON, C., THADANI, U., SINGLETON, W., TAYLOR, S.H.: Comparison of antihypertensive activity of beta-blocking drugs during chronic treatment. Brit. med. J. *1976/II*, 7
3 DAVIDSON, C., THADANI, U., TAYLOR, S.H., HESS, H., RIESS, W.: Pharmacological studies with slow-release formulations of oxprenolol in man. Europ. J. clin. Pharmacol. (printing)
4 GREEN, K.G., et al.: Improvement in prognosis of myocardial infarction by long-term beta-adrenoreceptor blockade using practolol. A multicentre international study. Brit. med. J. *1975/III*, 735
5 LAMBERT, D.M.D.: Long-term survival on beta-receptor-blocking drugs in general practice – a three-year prospective study. In Burley, D.M., et al. (Editors): Hypertension – its nature and treatment, Int. Symp., Malta 1974, p. 283 (CIBA Horsham, England, 1975)
6 MAROKO, P.R., KJEKSHUS, J.K., SOBEL, B.E., WATANABE, T., COVELL, J.W., ROSS, J., Jr., BRAUNWALD, E.: Factors influencing infarct size following experimental coronary artery occlusions. Circulation *43*, 67 (1971)
7 MARSHALL, A.J., ROBERTS, C.J.C., BARRITT, D.W.: Raynaud's phenomenon as side effect of beta-blockers in hypertension. Brit. med. J. *1976/I*, 1498
8 RIVIER, J.-L.: Beta-blockers in the treatment of angina pectoris: the prevention of myocardial infarction, loc. cit.[1], p. 52
9 SINGLETON, W., DAVIDSON, C., THADANI, U., TAYLOR, S.H.: Comparison of the immediate antihypertensive effectiveness of beta-adrenoceptor antagonists – a dose-response study. Clin. Sci. molec. Med. *48*, 18 (1975)
10 TAYLOR, S.H., MEERAN, M.K.: Different effects of adrenergic beta-receptor blockade on heart rate response to mental stress, catecholamines, and exercise. Brit. med. J. *1973/IV*, 257
11 TAYLOR, S.H., THADANI, U.: Oxprenolol in angina. Brit. J. Pharmacol. (printing)
12 TAYLOR, S.H., THADANI, U., DAVIDSON, C., SINGLETON, W., MYINT, S.: A comparative study of the activity of beta-adrenoceptor antagonists in man. Int. J. clin. Pharmacol. Biopharm. *12*, 305 (1975)
13 TAYLOR, S.H., THADANI, U., WATT, S.J., GOLDSTRAW, P., HESS, H., RIESS, W.: Studies with slow-release oxprenolol. II. Acute and chronic administration to patients with exercise-induced angina pectoris. In Judd, L. (Editor): Topics in cardiovascular disease, Int. Symp., Basle 1976, p. 131 (CIBA Horsham, England, 1977)
14 THADANI, U., DAVIDSON, C., TAYLOR, S.H.: Comparison of the antianginal activity of beta-blocking drugs. (In preparation)
15 WILHELMSSON, C., VEDIN, A., WILHELMSEN, L.: Influence of beta-blockers on the incidence of re-infarction and sudden death after myocardial infarction, loc. cit.[1], p. 58
16 ZANCHETTI, A.: Beta-blockers and cardiac involvement in hypertension, loc. cit.[1], p. 28

Summary of concluding discussion

Prof. Gross began by explaining that prior to the symposium Dr. Garnier, Dr. Matković, and Dr. Sarnoff had each been invited to prepare a brief contribution to the concluding discussion, which he now called upon them to present.

Dr. B. Garnier

For the many clinicians and specialists attending this symposium it may perhaps be of interest to hear something about the experience which a specialist in internal medicine has acquired with beta-blockers in the course of everyday practice. Hypertension and angina pectoris are the two main indications in which I myself employ these drugs. Summarised in Table 1 are the results I have obtained in 33 hypertensives whom I treated with oxprenolol (®Trasicor) plus diuretics. These results are evidently quite comparable with those that have been reported in large-scale studies such as the one published by Forrest*; and the picture that emerges from them would have been just the same if I had also included the additional cases of hypertension which I have started treating during the past few months. All 33 patients reviewed in Table 1 are still receiving treatment from me with a combination of oxprenolol and a diuretic.

From the practical standpoint, the fact that this treatment presents no difficulties is if anything almost more important than the quality of the results it yields. In other words, one of the main reasons why the prescription of a beta-blocker proves a satisfactory solution for the practising physician is because he knows that his patient will rarely, if ever, complain of side effects. This lack of side effects, moreover, also encourages patient compliance. The difference between beta-blockers and conventional antihypertensives was forcefully brought home to me when a specialist in hypertension whom I

Table 1. Data on, and results obtained in, 33 hypertensives treated in private practice with oxprenolol plus a diuretic. Included for purposes of comparison are figures relating to the average fall in blood pressure reported by Forrest in a general practice study in which the same type of regimen was employed.

33 hypertensives treated by the author	Age groups (years) and number of patients per group	< 40 5	40–49 7	50–59 11	60–69 8	> 69 2
	W.H.O. Stage and number of patients per stage	I 13	II 14	III 4	Renal 2	
	Blood pressure (systolic/diastolic)		Before treatment 184/113 mm.Hg		Under treatment 152/97 mm.Hg	
	2,300 hypertensives reported on by Forrest Blood pressure (systolic/diastolic)		Before treatment 184/108 mm.Hg		Under treatment 154/90 mm.Hg	
Treatment with oxprenolol plus a diuretic	Patients aged up to 50 years N = 15 Before treatment After treatment 172/113 mm.Hg 151/100 mm.Hg ⊿Blood pressure 21/13 mm.Hg			Patients aged over 50 years N = 18 Before treatment After treatment 193/113 mm.Hg 153/94 mm.Hg ⊿Blood pressure 40/19 mm.Hg		

* Forrest, W.A.: Oxprenolol and a thiazide diuretic together in the treatment of essential hypertension – a large general practice study. Brit. J. clin. Pract. 27, 331 (1973)

know, and who happens to suffer from high blood pressure himself, remarked: "Only since I started taking a beta-blocker have I realised how miserable I felt on diuretics and that reserpine had been making me depressive."

The patients for whom I have prescribed oxprenolol as treatment for angina pectoris have been fewer in number, but the results have been equally gratifying. Here, I have been particularly impressed not only by the drug's good symptomatic effect, but also by the change which it appears to produce in the course of the illness. Allow me to quote one example by way of illustration.

A 40-year-old crane operator – an obese patient with elevated blood uric acid levels – was subjected to coronary arteriography because he had been complaining of very severe angina pectoris. The right coronary artery was found to be obstructed, and there was also marked stenosis of the two left branches. The patient was therefore urgently advised to undergo a bypass operation. Since he refused to submit to surgery, he was placed on treatment with oxprenolol, prescribed in a dosage of 320–480 mg. daily. In response to this medication his condition improved and he was gradually able to resume work. He no longer experiences any pain in the mornings when he makes the 35 to 50 foot climb to the cabin of his crane, and in the evenings he returns home from work still free from pain. I find it difficult to believe that such a result can be attributable purely and simply to a symptomatic improvement.

In this connection I should like to suggest another possible reason for the cardioprotective effect of the beta-blockers – a reason to which no reference has yet been made during this symposium, namely, a change in cardiac mechanics. We know that under the influence of a beta-blocker cardiac contractions occur, as it were, more gently and smoothly. It seems to me reasonable to assume that this also exposes the coronary arteries to less mechanical strain, with the result that in the region of sclerotic foci, for example, there is less likelihood of endothelial tears, microhaemorrhages, or oedema occurring, which could aggravate existing coronary arteriosclerosis and possibly lead to an infarction or even to sudden death. This effect – in addition to the mechanisms already mentioned by Prof. BRUNNER – might also account for the favourable influence which beta-blockers exert on the clinical course of coronary heart disease.

Finally, I should like to endorse what Dr. TAYLOR said in his paper about the slow-release formulation that has been developed for oxprenolol. This new preparation introduced under the name of ®Slow-Trasicor, though perhaps of less importance for the hospital physician, strikes me as being especially valuable for use in domiciliary practice. I have had an opportunity of giving it an initial trial in hypertensives and patients with angina pectoris, for whom I have been prescribing it in a single daily dose of 1–2 tablets of 160 mg. to be taken in the morning. This dosage has kept the blood pressure of the hypertensive patients under good control, and in some cases the control achieved has been better than in response to the conventional oxprenolol formulation. In patients with angina pectoris its effect persists for at least 12–14 hours; in other words, the protection afforded by a single morning dose lasts till bedtime and also covers the patient's evening session of television. This new formulation has in fact simplified treatment even further; the number of tablets that have to be taken has been reduced, and there is hence less risk of the patient's forgetting to take a dose. I believe that, from the patient's standpoint, these are advantages which it would be impossible to exaggerate.

Dr. Z. MATKOVIĆ
Given the fact that the number of beta-blockers available is steadily increasing and that these various drugs, though rarely presenting any radical differences as regards

Table 2. Results obtained in an open, non-comparative clinical trial with oxprenolol, administered as monotherapy, in patients with essential hypertension of Grades 1–2.

Patients	N = 26 (♂ = 17, ♀ = 9)
Age	$\bar{x}$ = 42 years
Height	$\bar{x}$ = 170 cm.
Weight	$\bar{x}$ = 76 kg. (75 kg. after treatment)
Dosage of oxprenolol	2 tablets of 80 mg. twice daily (i.e. mornings and evenings) for five weeks

Mean blood pressure values (mm.Hg):		
Before treatment	170/110	
After 1 week of treatment	165/105	Side effects: two patients complained of
After 2 weeks of treatment	155/100	a bitter taste in the mouth.
After 3 weeks of treatment	155/100	No pathological changes were observed
After 4 weeks of treatment	150/100	in any of the biochemical or haematolog-
After 5 weeks of treatment	155/100	ical parameters.

their pharmacotherapeutic advantages or disadvantages, do nevertheless exhibit to a greater or lesser degree certain differing pharmacodynamic and pharmacokinetic properties, it is necessary – in countries where the introduction of new drugs is subject to statutory regulations* – to carry out clinico-pharmacological studies with these preparations, not only in order to promote the acquisition of further knowledge and experience but also because the performance of such studies is required by law.

The work we are currently engaged upon in this field** is primarily concerned with the elaboration of exact pharmacotherapeutic data, the testing of new beta-blockers in conformity with the relevant and accepted clinico-pharmacological principles, and the search for pharmacodynamic parameters of use in facilitating a differential assessment of these preparations.

Because the indications for beta-blockers, which had previously been reserved for the treatment of disturbances of cardiac rhythm and angina pectoris, have now been extended to include hypertension as well, and because we were also faced with the introduction of tablets containing a larger dose of active substance, we recently undertook an open, non-comparative clinical trial in which non-hospitalised cases of essential hypertension received monotherapy with oxprenolol in the form of tablets of 80 mg. The results obtained in this trial (an unpublished study carried out in cooperation with GOLEŠ, BENKOVIĆ, and WIESNER), which are outlined in Table 2, confirm findings reported by other authors. In the quest for additional parameters by which to assess the effects of this new antihypertensive treatment with beta-blockers, we also discovered that in patients suffering from essential hypertension the plasma concentrations of cyclic adenosine monophosphate are higher than in healthy controls, the difference being statistically significant***.

* Zakon o Stavljanju Lijekova u Promet: Službeni list SFRJ 6, p. 129 et sqq. (8.2.1973)
** ROSANDIĆ, D., MATKOVIĆ, Z., ĆUDIĆ, M.: Statistički prikaz farmakoterapije hospitaliziranih bolesnika primjenom određenih kibernetičkih metoda u bolnici "Dr. Josip Kajfeš" u Zagrebu za prva polugodišta 1974. 1 (1975). Summaries of papers presented at the IVth Meeting of Croatian Specialists in Internal Medicine, Split, 13th–15th November 1975, 8–9
*** RONČEVIĆ, T., MOLNAR, V., MATKOVIĆ, Z., BATES, M., VULETIĆ, J., VERČKO, K.: Ispitivanje cikličkog adenozin-monofosfata u urinu i plazmi kod arterijske hipertenzije. Summaries of papers presented at the XVIIth Meeting of Croatian and Slovenian Specialists in Internal Medicine, Karlovac, 3rd–5th June 1976, 21

Dr. S. J. Sarnoff

I should like to draw attention to another possible means by which – in one very specific situation – beta-blockers may, as it were, exert an indirect cardioprotective effect. In his paper, Prof. Brunner mentioned that acceleration of the heart rate steps up the oxygen requirement of the heart muscle and so increases the risk of myocardial ischaemia. This is certainly true. On the other hand, one common cause of death in the acute phase of myocardial infarction is hypotension, which is quite often due simply to severe bradycardia and is therefore in most cases relatively easy to correct.

Allow me to quote an example in point. In one of our patients we recorded a heart rate of 30 beats/minute 15 minutes after the onset of violent precordial pain. The patient was confused and exhibited signs of very extensive myocardial ischaemia. In response to an intravenous injection of 1.2 mg. atropine, the heart rate increased, an adequate coronary perfusion pressure was restored, and the myocardial ischaemia was thus kept more or less under control.

Injection of atropine is a simple and generally effective measure, but it does have certain disadvantages. A frequent side effect is severe tachycardia, which, together with an increase in myocardial contractility, causes a marked rise in the oxygen consumption of the heart muscle. This undesirable reaction can be prevented by administering a beta-blocker – as confirmed by the findings we obtained in a series of patients in whom we compared the heart rate following an intravenous injection of 1.5 mg. atropine with and without prior oral administration of a standard dose of propranolol. Whereas without prior beta-blockade five of the nine patients responded to the atropine injection with an acceleration of the heart rate to 100 beats/minute or more (i.e. up to 126 beats/minute), when the injection was repeated after a dose of propranolol had been given the heart rate in no case exceeded 100 beats/minute and in only one patient did it reach 90 beats/minute.

In a further series of 41 patients in whom atropine was injected after propranolol had first been administered, the heart rate rose above 90 beats/minute in only three cases, and in none of these three cases did it exceed 100 beats/minute.

These findings indicate that the effect of atropine in combating severe bradycardia can be more advantageously exploited if the patient is pre-treated with a beta-blocker, which exerts a genuine protective action by preventing an overshooting acceleration of the heart rate.

Prof. Gross, having thanked the three previous speakers for their prepared contributions, now declared the concluding discussion open to the floor.

Dr. Delius said he was very interested in Dr. Sarnoff's comments on the problem of severe bradycardia in patients with an acute myocardial infarction; in this connection he wished to ask Dr. Wilhelmsson whether he had been confronted with this problem in those of his patients who had had a re-infarction while under treatment with a beta-blocker.

Dr. Wilhelmsson replied that by and large bradycardia had not proved a problem either with alprenolol (®Gubernal) or with metoprolol (®Lopresor), which he had recently also been prescribing. Only one patient had had to be withdrawn from the study because his heart rate had dropped to below 50 beats/minute. But this did not apply to the acute phase of infarction.

Dr. Békes enquired whether Dr. Wilhelmsson had any observations to make on the so-called beta-blocker withdrawal syndrome.

Dr. WILHELMSSON explained that the studies carried out by his team had been so designed that treatment with the beta-blocker was withdrawn after two years, the withdrawal being effected without gradually reducing the dosage. During the first three months after this abrupt discontinuation of the medication, no deaths had occurred, there had been no re-infarctions, and no other signs or symptoms had been encountered which could have been interpreted as evidence of a withdrawal syndrome. Dr. WILHELMSSON admitted, however, that complications could conceivably arise in, for example, cases of unstable angina pectoris in which the beta-blocker had been withdrawn, say, in the "pre-infarction phase".

Prof. GROSS asked whether any member of the audience had ever observed such a withdrawal syndrome. He concluded from the silence which greeted his question that apparently no one had.

Dr. SCHLESINGER asked Dr. WILHELMSSON if he could please comment on the incidence of heart failure among his patients.

Dr. WILHELMSSON stated in reply that some patients had been excluded at the very outset because of cardiac decompensation. During the study, heart failure had developed in eight of the patients in the beta-blocker group and in eight of those in the placebo group, but it invariably responded satisfactorily to routine countermeasures.

Dr. FUCCELLA objected to the definition of an ideal beta-blocker as given in Dr. TAYLOR's paper. The argument that such a beta-blocker should not be cardioselective conflicted with the desirable aim of minimising the influence exerted on the bronchi, the postulate that the drug should possess intrinsic sympathomimetic activity was incompatible with the goal of maximum suitability for use in all indications for beta-blockers, and, finally, an ideal beta-blocker should surely be one that also exhibited as little membrane-stabilising activity as possible.

Dr. TAYLOR replied that intrinsic sympathomimetic activity also had a beneficial influence with regard to airway resistance; he added that extensive investigations in patients with hypertension and angina pectoris – which were by far the most important indications for beta-blockers – had failed to disclose any differences in the clinical effectiveness of cardioselective and non-cardioselective beta-blockers with or without intrinsic sympathomimetic activity. As for the membrane-stabilising effect of beta-blockers, this effect was so minimal in response to doses of the size normally employed that it could safely be ignored.

Dr. POOLE-WILSON felt that the precise significance of the cardioprotective action of the beta-blockers should be more accurately defined. Did this action have purely negative advantages, i.e. was it confined to reducing the risk of myocardial infarction and sudden death, or could beta-blockers also be expected to produce more positive effects when administered on a long-term basis? Were these drugs to be employed as a form of general cardiovascular prophylaxis, or should one limit their use to certain specific types of patient?

Prof. GROSS reminded the audience that the term "cardioprotection" as applied to the action of beta-blockers was a simplified formulation the exact implications of which required to be defined. He asked if Dr. TAYLOR would care to add a final comment.

Dr. TAYLOR said he would be in favour of resorting to a beta-blocker in patients at risk, i.e. whenever any event or events occurred which threatened to accelerate the progress of a cardiovascular disease, and that in such cases his aim in instituting treatment with a beta-blocker would be to prevent cardiovascular complications.

Closing remarks

by F. Gross*

At the end of our symposium, it may be appropriate to remind you of its title: "The cardioprotective action of beta-blockers – facts and theories". We have heard quite a few facts, some confirmative, others new, and various theories have been aired in the presentations and in the complementary discussions. In my introductory remarks I drew attention to the necessity of setting up standards when we talk about "cardioprotection" and of defining the criteria for a protective activity of beta-blockers. I should like to add that we must also distinguish between acute events and chronic states when we try to influence them by blocking beta-adrenergic receptors. A chronic effect which, as shown by Prof. Guazzi, remains stable in the long-term treatment of hypertension, or one which serves to afford protection against re-infarction, is different from what can be achieved in preventing sudden death or in preserving myocardial tissue after infarction. I must say that I was most impressed by the data that Dr. Wilhelmsson presented on the Gothenburg studies and on the cooperative study by Green and his associates, by Stewart's results, which Prof. Zanchetti mentioned, and by the observations made by Dr. Taylor. Adding all these reports to those already familiar to us, I think that Dr. Somerville will have to use not only his five fingers but a few more in order to count the proofs available today for the cardioprotective activity of the beta-blockers!

Of course, we need more data, and in this respect our symposium has fulfilled a trigger function. On the other hand, in view of the findings reported by Dr. Wilhelmsson, I doubt whether it would be ethically justifiable to organise another double-blind trial with a placebo group. But there are drugs other than beta-blockers which could be used in patients with acute myocardial infarction, and doctors who have the opportunity to compare treatment schedules with and without beta-blockers should do so.

I said at the beginning of this symposium that the term "cardioprotection" is used merely for the sake of simplicity; we have to realise what we mean when we apply it, and we have to beware that it does not become just a pseudo-scientific neologism. Properly employed, however, it is a helpful term, which aptly describes what we should like to achieve. There is certainly a need for cardioprotection – i.e. to limit the extent of myocardial necrosis, to protect patients against sudden death, to antagonise the effects of exaggerated catecholamine release, and to prevent thrombosis and re-infarction. If this can be accomplished, at least to a certain degree, with the help of beta-blockers, then we shall have taken an important step forward in the treatment of heart diseases.

No drug, and no group of drugs, is perfect, and I doubt whether the ideal cardioprotective beta-blocker that Dr. Taylor dreams of will ever be found. The more active a drug is, the more carefully it has to be applied, and the more skill and knowledge are necessary for its optimum use. We may be thankful that the beta-

---

* Pharmakologisches Institut der Universität, Heidelberg, German Federal Republic.

blockers have such a wide margin of safety and are in general quite well tolerated; they would be a nightmare if they had a therapeutic range as narrow as that of digitalis.

Allow me to end with a word of thanks. Firstly, I should like to express my gratitude to all the scheduled speakers for their excellent papers and contributions to the discussions, as well as to all those who stimulated the meeting by their comments from the floor. My thanks also go to the organisers of this symposium and to their technical staff – the projectionist, who showed all the slides not only in the correct order, but also the right way up, and the interpreters, who had to do a hard job under cramped conditions. Finally, let me thank CIBA-GEIGY, who sponsored this symposium. Today it is fashionable to criticise the pharmaceutical industry – and some of the criticism is certainly not unjustified – but we should also give credit to those drug companies whose scientific achievements (it was the pharmaceutical industry that developed beta-blockers) and whose material contributions make it possible to hold meetings like the one we have attended this morning. We should bear in mind that, without the support of the pharmaceutical industry, many of us could not have performed some of our research activities and might not be able to participate in gatherings such as the congress we are enjoying this week. This help and encouragement which most, if not all, of us have already experienced should for once be gratefully acknowledged.